Betalains: Biomolecular Aspects

Erum Akbar Hussain • Zubi Sadiq
Muhammad Zia-Ul-Haq

Betalains: Biomolecular Aspects

Springer

Erum Akbar Hussain
Lahore College for Women University
Lahore, Pakistan

Zubi Sadiq
Lahore College for Women University
Lahore, Pakistan

Muhammad Zia-Ul-Haq
Lahore College for Women University
Lahore, Pakistan

ISBN 978-3-319-95623-7 ISBN 978-3-319-95624-4 (eBook)
https://doi.org/10.1007/978-3-319-95624-4

Library of Congress Control Number: 2018949117

This Springer imprint is published by the registered company Springer Nature Switzerland AG
The registered company address is: Gewerbestrasse 11, 6330 Cham, Switzerland

Preface

Besides the health-promoting effects of betalains and adding splashes of color to the world that we live in, betalains are still not considered "nutrients." Nonetheless, betalains are receiving increased attention as a group of phytochemicals important for optimal health. Betalains are a significant source of important antioxidants, and plant-based diets are consumed by populations most at risk for deficiency of antioxidants. Providing information about health-promoting and disease-lowering effects of betalains may decrease the prevalence of various diseases. The overarching goal of this book is to convince the reader that betalains can contribute to overall health and well-being in addition to their well-known color-imparting function.

The inspiration for *Betalains: Biomolecular Aspects* was to provide health-care and nutrition professionals and medical, graduate, and senior undergraduate students with a resource of up-to-date information on betalains. Betalains are ubiquitous pigments found mainly in plants and fungi and are considered as major contributors to the health benefits associated with diets rich in fruits and vegetables. The major objective of this comprehensive book is to review the growing evidence that betalains are bioactive molecules that can be of value to many aspects of health.

The different chapters of this book complement each other and provide distinct areas to be used for teaching. The first chapter, "Introduction," provides essential background for all readers on dietary sources of betalains and introductory material on various aspects of betalains that can be used in graduate-level instruction. The second chapter, "Sources of Betalains," may be used by practitioners and for senior undergraduate, graduate, and medical school-level courses on the importance of betalains in human health and development. Health-care and nutrition professionals will find this section most informative as they advise patients and clients.

The third chapter, "Chemistry of Betalains," is an up-to-date comprehensive review of the science behind the active molecules of betalains. It is of great importance for chemistry, pharmacy, and health graduates who have to rationalize chemical aspects with bioactivities of these colorful pigments. The fourth chapter, "Biosynthesis of Betalains," elaborates various pathways of synthesis of betalains in plants and describes research concerning the metabolizing enzymes and sites of their activities as well as the mechanisms of transport of betalains in plants.

The fifth chapter, "Role of Betalain in Human Health," examines the role of betalains in human health and begins with the importance of adequate betalains intake to assure personal health and well-being. This chapter will serve the reader as the most authoritative resource in the field to date.

The sixth chapter details "Bioactivities of Betalains." Special focus is given on antioxidant and singlet oxygen quenching reactions performed by betalains and their biological significance in promoting health. This chapter reviews the data from relevant clinical studies so that the current and past findings can be placed in the proper perspective, especially with regard to the potential for new data to suggest that certain betalains may have benefit as anticarcinogenic agents. Betalains, have the capacity to impart colors. The seventh chapter, "Betalains as Colorants and Pigments," highlights the coloring characteristics of betalains.

The eighth chapter, "Analysis of Betalains," reviews progress in analysis of betalains. All techniques starting from simple like paper chromatography till advanced like spectroscopy have been described including pros and cons of every technique and method.

The next chapter, "Bioavailability of Betalains," summarizes the impact of other dietary components on the metabolism of betalains. In addition to dietary factors affecting betalains absorption and metabolism, host factors also can greatly affect betalains bioavailability. Host factors can influence the ability to absorb, convert, and metabolize dietary betalains. Factors such as gender, body fat, and genetic variation play an important role in this process. The last chapter, "Processing of Betalains," examines the effects of various processing techniques on chemistry and biology of betalains. This chapter is useful to the nutrition community as well as for health professionals who have to answer client or patient questions about this area of clinical research.

It is hoped that this book will ignite scientists, practitioners, and students to evaluate their work and endeavors in the scheme of global public health. Although betalains are not currently considered essential nutrients, as we move from prevention of nutrient deficiency to supporting optimal human health and prevention of disease, evidence presented in this book should compel the reader to contemplate what truly defines a nutrient.

Lahore, Pakistan

Erum Akbar Hussain
Zubi Sadiq
Muhammad Zia-Ul-Haq

Contents

Chapter 1
Introduction

1.1 Historical Aspects

Man is always fascinated with colors, especially when edibles are under consideration; we are all captivated with colorful foods. Betalain is one of the important natural pigments of the food industry and safe from the health point of view. Due to inextensive research in chemistry, biosynthesis and ecophysiological factors affecting betalain accumulation and evocation in situ/ex situ for its improved production were for the first time calculated by its annual production potential estimation, and relevant future study was attempted [1]. Betalains are named as chromo-alkaloids that are polar, hydrophilic nitrogenous pigments which mainly exist in most plants of Caryophyllales order [2, 3]. It is derived from *Beta vulgaris* from which its extraction was done for the first time and well recognized as a chief natural source. Beetroot is the main part of plant which has enormous quantity of betalain than any other part. The presence of carboxylic acid is responsible for the acidic nature of this important bioactive molecule, which is why it is not included in alkaloids [3]. The earliest chemically identified betalains were thought to be anthocyanins till 1957 or nitrogenous anthocyanins more incisively [1] because the biological functions of anthocyanins were replaced by these nitrogenous compounds in plants [4]. This term incorrectly suggested structural resemblance between the two pigment classes: both betaxanthine and anthocyanin [5].

The structure elucidation was made possible in the recent years due to the advancement in spectroscopic techniques particularly the nuclear magnetic resonance technique. Within the last era, ample characterization of some presumed betalains as well as their biosynthetic pathways was described that was helpful in better understanding of betalain structure. Biogenetic and structural evidences showed that the term "betalains" was introduced by Mabry and Dreiding. By spectroscopic techniques, betaxanthins with some other amines and amino acids in addition to several betanidin conjugates (glycosides and acylglycosides) were recognized afterward [6–8].

E. Akbar Hussain et al., *Betalains: Biomolecular Aspects*,
https://doi.org/10.1007/978-3-319-95624-4_1

Betalains are nitrogenous pigments which are subdivided into betacyanins (red-violet) and betaxanthins (yellow-orange) (Fig. 1.1). Phenylalanine-based, red/violet secondary metabolic anthocyanin pigments are produced by all existing flowering plants except for a few families of the order Caryophyllales in which dissimilar violet/red/yellow beet pigments or betalain is produced.

Betalain and anthocyanin pigments are equally exclusive; no taxa are known to produce both pigments [1]. Betalains replace the anthocyanin physically and functionally, in all biological contexts. They are free oxygen radical scavengers which are nutritionally beneficial in ways similar to the phenylpropanoids such as the anthocyanins [4]. All the major crop families produce anthocyanins. The betalain-producing families include such crops that are widely grown and associated with various important agricultural economic systems, such as beets, Swiss chard, spinach, *Amaranthus*, *Chenopodium quinoa*, and prickly pear. Evidences by crystallization of betanin, hydrolysis of betanin to betanidin, and subsequent report on indicaxanthin isolation show that they are a different set of pigments having a 1,7-diazaheptamethin system which is responsible for their chroma.

Unique biosynthetic pathway toward betanin and indicaxanthin through DOPA incorporation further reinforces this concept. Caryophyllales-specific occurrence of betalains shows chemotaxonomic pertinence to secondary metabolites of plants. Betalains are still a secret due to evolution in plant and net gain/loss of anthocyanins. Clarification of evolutionary mechanisms leading to the communal segregation of betalain and anthocyanin pathways in flowering plants requires more extensive molecular studies [6]. In case of leaf margin coloration, the pigment responsible for reddening/purpling is presumed to be anthocyanin, and there are no reports of leaf margin reddening by carotenoids or betalains [9].

1.2 General Characteristics

Betalains show a large diversity in structures due to acylation and glycosylation. In plants, conjugates of the chromophore betalamic kingdom, betanin (betanidin-5-O-β-glucoside) is the most common betacyanin acid [7], derived by an oxidative 4,5-extradiol ring-opening mechanism from 3-(3,4 dihydroxyphenyl)alanine (DOPA) [10].

Due to antioxidative, anti-inflammatory, and anticarcinogenic properties, betalains are of additional interest. They are used for coloring dairy products, meat, and frozen desserts. Like anthocyanins, betalains act as optical attractors for pollinators and for distribution of seeds. Their synthesis induced in *Mesembryanthemum*

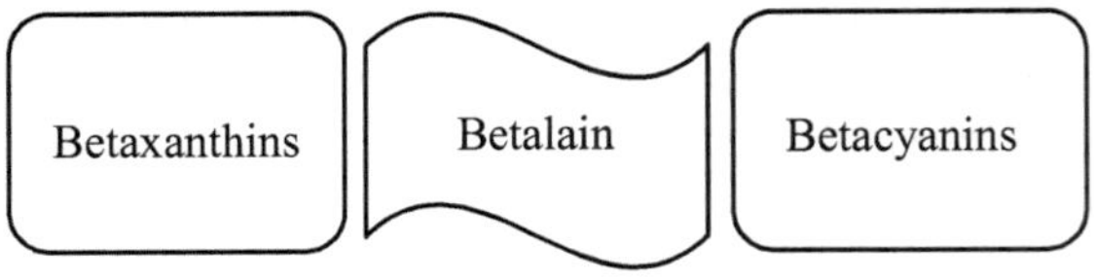

Fig. 1.1 Classification of betalain

crystallinum L. (ice plant) by UV radiation and in *Beta vulgaris* L. (red beet) by viral infection reveals that betalains also have radioprotective and antiviral activity showing its recent increase in interest due to their desirable chemical, medical, and pharmacological properties; i.e., they are chemically stable over a wide range of pH as compared to anthocyanins, possessing anticarcinogenic and anti-inflammatory activities, and they serve as strong antioxidants [11].

The natural pigments that are safe for human health are quiet scarce to be used in food. Pigments are under strict regulations, because the US Food and Drug Administration (FDA) considers the pigments as additives so the endorsement of new sources is challenging in terms of its consumption for human. The better use of pigments will be improved by an adequate empathy of the actual sources and preferred if proved safe and derived from natural source [3]. Betalains have many sources as mentioned in Fig. 1.2. In the order Caryophyllales, plants Molluginaceae and Caryophyllaceae are exceptions which produce anthocyanin, while betalains are produced in roots as well as flowers in *Basidiomycetes* (i.e., *Hygrocybe conica, A. muscaria*) [7].

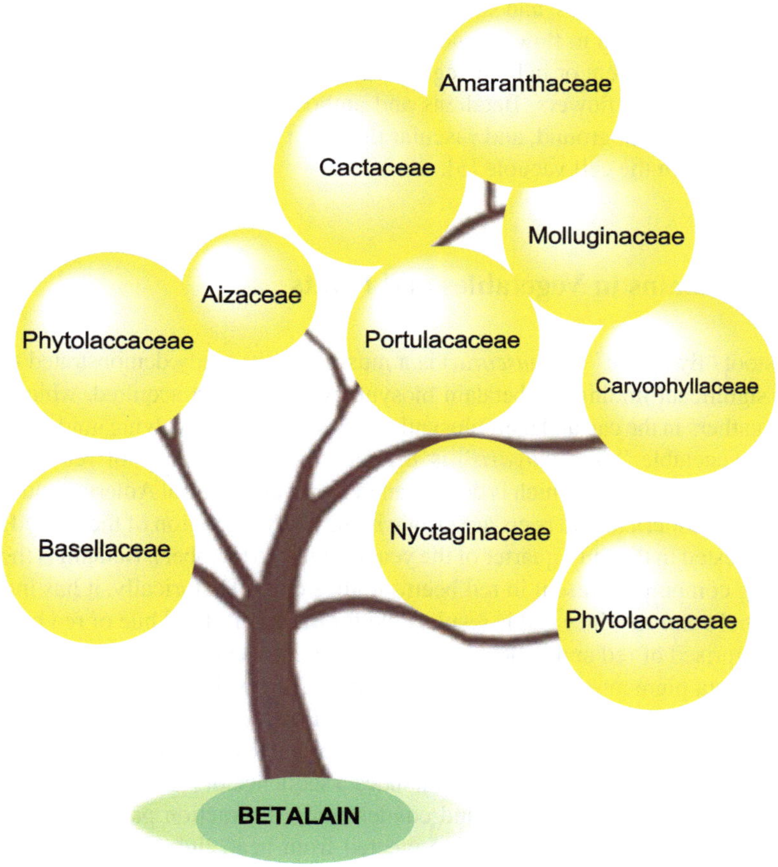

Fig. 1.2 Sources of betalain

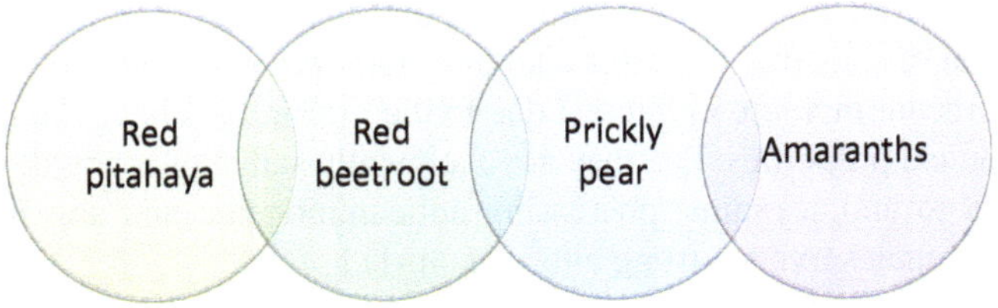

Fig. 1.3 Chief sources of betalains

The existing sources of betalains such as amaranths, prickly pear, and red pitaya other than red beetroot should be grown in appropriate quantities to make sure they are manufactured at a large scale to be used in the food industry. Their growth can be enhanced via breeding and high-yielding biotechnological tools, such as plant tissue cultures and genetic engineering [14] (Fig. 1.3).

Betalain occurrence is restricted in plant families belonging to the Caryophyllales and higher fungi, where seeds, fruits, flowers, leaves, stems, and/or roots are the specific parts of plants which contain betalain pigment. The distribution of betalains in plant parts, their colors, and sources are mentioned in Table 1.1. In cacti, betalain synthesis is restricted to flowers and fruits (reproductive organ) from a wide range of natural environment or in both vegetative and reproductive structures of ice plant such as leaves and flowers. Betalains and anthocyanins share similar histological locations in dermal, ground, and vascular tissues of vegetative organs and are stored as glycosides in the cell vacuole [5].

1.3 Betalains in Vegetables and Fruits

Toadstool "fly agaric" (*A. muscaria*) is a mushroom; its study demonstrated that the most significant findings of betalain biosynthesis have been acquired, which reveal that it gathers in the cap and their biosynthesis is subjected to evolving guidelines [15].

The vegetable "*Beta vulgaris*" is the most common source of red beet juice extracted from its roots which is cultivated in North and Central America and Britain [14]. North American red beets are grown in the Midwest region of the United States and harvested in the third quarter of the year (August to October). Betanin is the main coloring compound present in red beetroot juice color. Historically, it has imparted additional color to wines and is responsible for the color of red hue of red beet juice and comprised of red and yellow pigments known as betacyanins and betaxanthins, the magenta pigment and the yellow pigment, respectively. Red beetroot hues vary depending on betalains' source of extract. The dissemination of extracted pigments differs owing to aspects such as beetroot cultivar and extraction method.

A prevalent extraction technique includes a sequence of size reduction processes followed by hydraulic filtration and condensation. Extraction parameters are controlled in such a way that color is protected from heat, light, pH, and enzymatic degradation during the process.

Table 1.1 Distribution of betalains in plants

Structure of plant	Leaves and stems Bracts Flowers Roots Seeds Fruits
Color produced	A long range of colors Color ranges Yellow, orange, red, and pink Red-purple Red and yellow, among others Purple, red, and yellow
Sources	Teloxys spp. *Bougainvillea* spp. *Portulacaceae* plants and *Aizoaceae* Red beetroot *Amaranthus* spp. Prickly pear, *Opuntia stricta*

The contents of beetroot products depend upon the red beets' cultivar. The major yellow components are xanthin I and vulgaxanthin II and red components are betanidin and betanin (as well as their isomers). These are a unique source of phytonutrients. Vulgaxanthin and betanin are well-explored betalains obtained from red beets. Anti-inflammatory, antioxidant, and detoxification maintenance are associated with both of these. The detoxification maintenance by betalains comprises sustenance of certain important Stage 2 detox steps comprising glutathione. Although these pigments can be used in diet (such as chard or rhubarb stems), the betalain concentration in the bark and flesh of beets gives an unpredictably pronounced chance for health fitness.

The betanin concentration is decreasing in the following order:

Peel > crown > flesh

Red beet contains about 300–600 mg/kg betanin. Various factors affecting its content in betanin are cultivar, farming conditions, temperature during the growing season, soil fertility, soil moisture, storage temperature, etc. In the food industry red beet extracts containing betalain are allowed in the USA as natural food colorants [7] (Fig. 1.4).

Swiss chard scientifically known as *Beta vulgaris* is a large green leafy vegetable having a bushy and crispy stem that exists in red, white, or yellow fanlike leaves. Chard, spinach, and beets belong to the same family and have the familiar strong bitter and somewhat salty taste. It exhibits exceptional health-improving properties. It grows faster in June and August, but it is widely available for the whole year.

The term "greens" is widely referred to leafy vegetables like Swiss chard, mustard, cabbage, and collard greens possessing curly or smooth leaves, liable upon the diversity, and light-colored spine feature distributed end to end. It exhibits orange, yellow, red, or white variety in colors to the stem that is nearly two feet long. In the

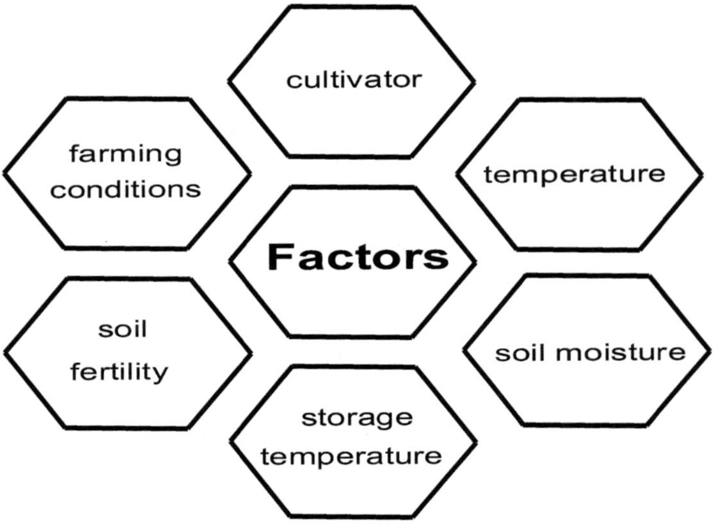

Fig. 1.4 Various factors affecting red beet content in betanin Swiss chard

market, occasionally, clusters of various colors are huddled and branded as "rainbow" chard. It is a regular biennial plant which is seeded in June, July, August, September, or October in Northern areas. Its garnering is a nonstop method due to the production of three or more harvested crops by a number of chard classes. It contains betalains, phyto-nutritional pigments such as betaxanthins (yellow) and betacyanins (red). Its royal purple-colored veins and stem exhibit different betacyanins such as isobetanidin, betanidin, isobetanin, and betanins. In stems and veins possessing yellow color, 19 different betaxanthin pigments are present such as triamine-betaxanthin, alanine-betaxanthin, histamine-betaxanthin, as well as 3-methoxytyramine-betaxanthin. Its biological profile includes anti-inflammatory, antioxidant, and detoxification support. Phytonutrients present in betalains can rapidly detoxify the gut due to its anti-inflammation and antioxidant properties. The nerve that is responsible for vision signaling is well protected by betalains.

Betalains are present in the fruits of prickly pear or cactus plant scientifically known as *Opuntia ficus-indica* L. mainly in purple variety as betacyanins and in orange variety as betaxanthins. The origin of cactus pear is from Mexico, but it also comes from Africa, Middle East, and Europe later on revealing extraordinary alteration to semiarid and arid type of weather in subtropical and tropical areas of the universe. It is self-generated in Italy and is cultured in Southern areas such as Calabria, Apulia, Sardinia, and Sicily.

Due to flavorsome properties, fruits of this plant are commercially significant. Usually, the fruit is taken fresh in July to October that is the seasoning period, but its shelf life has also been increased by various procedures developed by food technicians due to growing arcade request for health-encouraging food. Due to its nutritive and health-encouraging features, it has gained a considerable prominence. It is also associated with bioactive ingredients such as antioxidants like polyphenols, ascorbic acid, and betalains. Apart from this, antiproliferative, hepatoprotective, neuroprotective, anticancer, and antiulcer organic activities are also associated with it. It is also an excellent source of yellow and red dyes for food.

As stability range of betalains ranges from 4 to 7 pH, it is most favorable for non-acidic food dyeing. Furthermore, the existence of both betaxanthins and betacyanins at the same time produces a broad range of color.

Almost 50% of the total cactus pear is produced in Mexico that is the main worldwide producer. After that, Italy comes next that produces 78,000 tons and 7400 Ha. In Italy, 90% production is that of Sicily, whereas Apulia documented in 2013 the creation of 2650 tons from 320 Ha, chiefly developed in North Apulia (Foggia province) with particular crops that are spineless. In Italy, exhaustive plantations principally propagate the spineless yellow variety. There grows an equal sharing of the two colored fruits in Apulia, predominantly in the South Apulia (Salento peninsula), from bristly genotypes, habitually rising privately or in wild country parks, by means of an unlike equilibrium of betacyanin (purple-red) present in purple variety and betaxanthin (yellow-orange) present in the orange variety.

Opuntia stricta is a cactus species commonly named as erect prickly pear from South to North America and the Caribbean. It is a sprawling or erect shrub up to two meters in height, bearing light yellow flowers from February to June and producing purplish-red fruits. *Opuntia stricta* contains five times higher concentration of betalain than *Opuntia ficus-indica* and partly higher as in red beet. *Opuntia stricta* occurs naturally in the Southern region of the United States as well as the Bahamas, Bermuda, the Caribbean, Central America, northern Venezuela, and Ecuador. Many regions of the world are beautified with this medicinally important plant; these regions include Africa, Southern Europe, and Asia particularly in Sicily. *O. stricta* is believed as a persistent species in South Africa. In Australia, it has been the subject of one of the first effective biological control exercises against the moth *Cactoblastis cactorum*. Literature has highlighted it as "Weeds of National Significance," though it must be used under controlled conditions.

New Zealand has ice plant, one of its inhabitant plants, which is regarded as a key source in augmenting crop capacity to salinity. This plant has red pigmentation that is believed to be one of the responsible factors for its saline resistance whose variable amount depends on its remoteness from the coastline. Betalain, the red pigment, in the ice plant, was studied, while the incorporation of betalains in green-leaved ice plant makes the plant tolerant from saline conditions which functions as a defense for plant tissues contrary to free radical destruction triggered by salt and too much sunlight. This result specifies the potential of betalain to be used in emerging and breeding crops to enrich forbearance under saline state.

Amaranthus, a member of family *Amaranthaceae*, is widely cultivated in some areas of Asia and Africa and contains complex mixtures of betalain. Several species of *Amaranthus* are often considered as weeds. Yet, in many countries, it is also used as vegetables, cereals, as well as medicinal and ornamental plants. Interestingly, the cultivated species contain more betacyanin and higher biomass than wild ones. Based on its physical/chemical properties, its pigments show possible potential as food colorant and also strong antioxidant activity, so it is recommended as an alternative source of betalain. In China, a natural pigment of *Amaranthus tricolor* has been legally accepted as food ingredient (Hygienic Standards for Food Additives, GD2760-89). The cultivars and the growth stage affect the pigment concentration which shows that the cultivated genotypes have a potential to be commercially developed as natural pigment sources. More than 60 species of short-lived herbs have a cosmopolitan genus that is known as *Amaranthus*. Cultivation of some amaranths is done as ornamental plants, leafy vegetables, or grain, whereas other amaranths are cultivated in the form of weeds. *Amaranthus tricolor* contains less betacyanins in their tissues than *Amaranthus hybridus*, *Amaranthus cruentus*, and *Amaranthus caudatus* (grain amaranths) [4]. Three kinds of betaxanthins and 16 kinds of betacyanins have been identified and elucidated in this plant family [11].

Pitaya (*Hylocereus undatus*) is the member of *Cacteacae* family, instinctive of Mexico and Central and South America. It comprises of numerous species though *Hylocereus undatus* is superlative to be cultivated. Pitaya peel is reflected as a residue from utilization and treatment of fruit, and it is usually discarded. However, this remainder can be consumed as feedstock for the extraction of pigments, due to the presence of betalains which display substantial and attractive color. Beet (*Beta vulgaris*) is the chief source of commercial extraction of betalains that comprises of geosmin and *pyrazines* which are accountable for the loss of original flavor due to deterioration. The betalains can be used in foods without flavor, which are extracted from pitaya, contrasting the red beet, and it covers an extensive spectrum of color from yellow-orange of *Opuntia* to red-violet of *Hylocereus*.

Strawberry blite belongs to the edible group of yearly grown plant. Moreover, *Blitum capitatum* syn. *Chenopodium capitatum* is also known as blite goosefoot, strawberry spinach, and strawberry goosefoot. It can be found in North America but may also be found in Canada, New Zealand, and the United States. Strawberry blite also occurs in damp mountain valleys. Its fruits are bright red and small but pulpy and edible just like strawberries. Natives use its fruit extract as red dye. The fruits have small, lens-shaped seeds that are black and 1.2–0.7 mm long. Its edible portion must be used in moderation due to oxalates. Seeds are toxic when used excessively (Table 1.2).

1.4 Betalain: A Versatile Color Source

Betalains obtained from red beet are responsible for brilliant colors in flowers and fruits belonging to species of the order Caryophyllales, excluding Caryophyllaceae and Molluginaceae used as natural colorant. The mutual clannishness of Caryophyllales has raised appreciable taxonomic debate. Lack of understanding of betalain biosynthesis in terms of molecular biology and biochemistry has prohibited researchers from interpreting this taxonomic puzzle. Betalain color is beneficial because this is independent of pH and is more stable than the color obtained from anthocyanins [12]. They are not only responsible for the bright coloration of fruits and flowers but also of leaves and roots of plants in Caryophyllales order [13].

Table 1.2 Major betalain-containing plant sources [11]

Common name	Species	Betalains produced
Beetroot	*B. Vulgaris*	Indicaxanthin, vulgaxanthin (I, II), betanin, isobetanin, prebetanin, neobetanin
Swiss card	*B. vulgaris* (L.)	20 different betaxanthins 9 different betacyanins
Prickly pear	*Opuntia ficus-indica* (L.)	Minor api of uranosyl betacyanins, hylocerenin, and isohylocerenin
Erect prickly pear	*Opuntia stricta*	Betalanin, betacyanin
Ice plant	*Mesembryanthemum crystallinum*	Mesembryanthin, betacyanins
Amaranth	*Amaranthus* spp.	Amaranthine, isoamaranthine
Pitaya peel	*Hylocereus undatus*	Isobetanin, betanin, isophylo cactin, phylocactin, hylocerenin, isohylocerenin
Strawberry blite	*Blitum capitatum*	Betanin
Red goosefoot	*Chenopodium rubrum*	Amaranthin, vulgaxanthin (I, II), betanin, celosianin
Moss rose	*Portulaca grandiflora*	Dopaxanthin, portulacaxanthin II, vulgaxanthin I, indicaxanthin, miraxanthin V

Its color shows stability in pH range 3–7. This range is useful for *Mesembryanthemum crystallinum* and *Portulaca oleracea* which are betalainic plants, and their carbon assimilation mode is switched from C4 and C3, respectively, to crassulacean acid metabolism (CAM) when it experiences water stress. A shift from a relatively stable vacuolar pH to large periodic fluctuations is observed in these plants. In the plants that occur in drought conditions, a stable red color reveals irrespective of vacuolar pH and maintenance of photo protection.

In betalain synthesis, nitrogen requires an additional cost; two atoms of nitrogen per betalain are required. So, there's a need to balance this economy, in spite of keeping equality between both pigment classes; anthocyanins are more successful than betalains as defensive molecules [5].

1.5 Classification

In the class *Basidiomycetes* of fungi and within the Caryophyllales's suborder chenopodiaceae, betalains are restricted [11]. All betalains have the same basic structure in which R1 and R2 may be substituted with hydrogen and aromatic substituent. The colors of betalains are the result of their resonating double bonds [15]. A substituted dihydropyridine serving as the chromophore with a system of conjugated double bonds is the basic structure of betalains (Fig. 1.5). Betalamic acid forms immonium conjugate with amines or cyclo-3-(3,4-dihydroxyphenyl)-L-alanine (cyclo-DOPA) to form betalains (Fig. 1.6) [11].

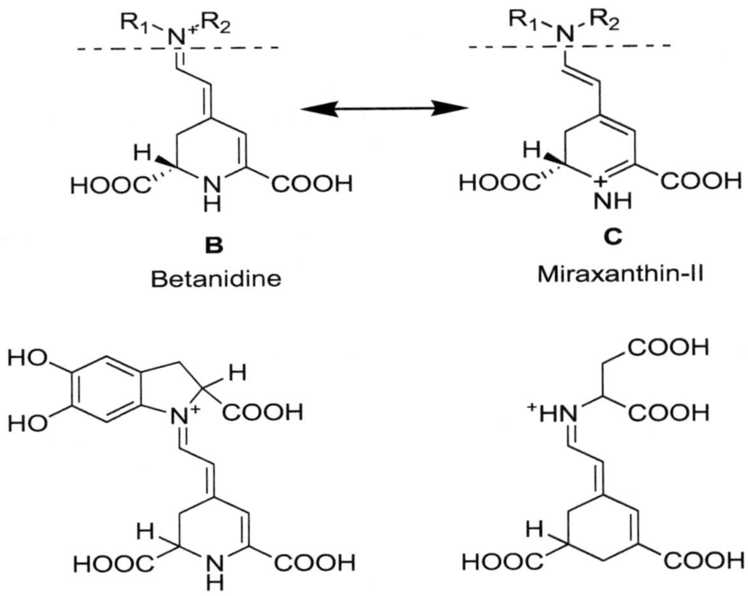

Fig. 1.5 Basic structure of betalain

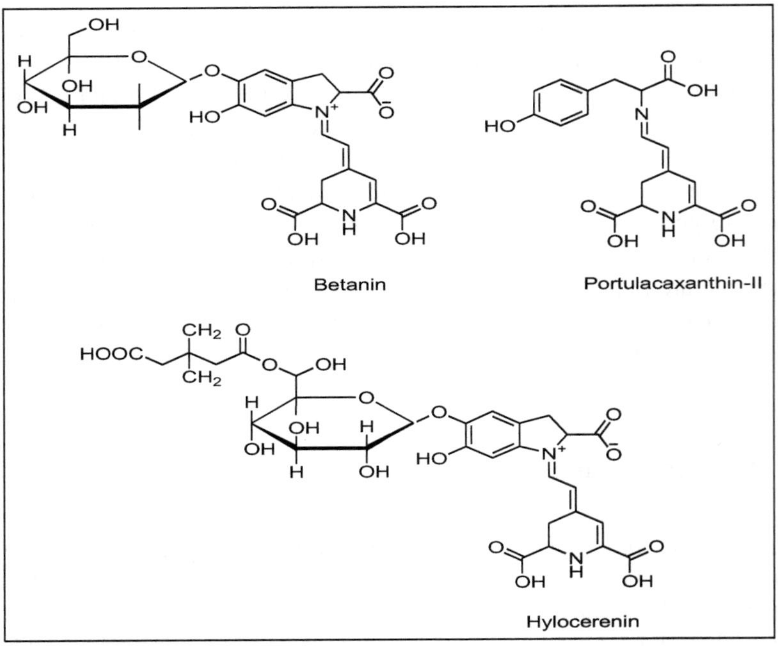

Fig. 1.6 Chemical structures of betalain

More than 100 betalains are classified in two groups, depending on ligand R1-N-R2 moieties [3] and based on their structural characteristics:

- Betaxanthins
- Betacyanins

A change in the absorption maximum from 540 nm (red-purple betacyanins such as betanidine) to 480 nm (yellow betaxanthins such as miraxanthin II) is observed when the basic structure is substituted with an aromatic nucleus [15].

The terms "betaxanthines," "betalaines," and "betacyanines" were used in some earlier papers. Fisher and Dreiding added the terminal letter "e," which is confirmed by nomenclature of IUPAC. These terms can also be used at present without the letter "e" [3].

Different amines and amino acids form immonium derivatives of betalamic acid to construct betaxanthins (from Latin beta, red beet, and Greek xanthos, yellow [3]). Cyclo-dihydroxyphenylalanine (cyclo-DOPA) condenses with betalamic acid to form betacyanins (kyanos, blue color [3]) [13].

On the basis of chemical structure, betaxanthins and betacyanins are classified. Feruloyl (which is an acyl group) and sugar (e.g., 5-O–D-glucose) moieties show variation in betacyanin structure, whereas a wide range of amino acids (e.g., tyrosine) and amines (e.g., glutamine) show conjugation with betaxanthin structure. Some well-studied betalains are shown in Table 1.3 [3].

There are four subgroups into which betacyanins are divided into: betanin, amaranthin, gomphrenin, and 2-descarboxybetanin type (Fig. 1.7).

Red betanin is a well-known betacyanin that gives red beet (*Beta vulgaris* ssp.) a typical red color, and betaxanthins are further classified into three subgroups (Fig. 1.8) [7].

B. vulgaris contains vulgaxanthin I which is formed by conjugation of glutamic acid at R2 group. In the same way, amines such as amino acids show conjugation with dihydropyridine moiety to form large numbers of betaxanthins. At positions 5 and 6, hydroxyl groups are present which are attached with different glycosyls such as glucose (most common moiety), sophorose, and rhamnose (occur less frequently) and acyl groups such as sinapic, sulfuric, 3-hydroxy-3-methylglutaric, caffeic, malonic, citric, and ferulic acids with the basic structure (betanidin is the most important, followed by isobetanidin, its C15 epimer) and form diversity in betacyanins. Examples of betalains are shown in Table 1.4 [15].

The more general classification of pigments can be understood under the following subheadings:

By the Origin of Pigments

Pigments are classified as natural, inorganic, or synthetic origin. Plants, fungi, microorganisms, and animals which are living organisms produce natural pigments. From laboratories synthetic pigments are obtained. Inorganic pigments are synthesized and are also found in nature, whereas organic compounds can be natural and synthetic

Table 1.3 Some fully studied betalains

Betalain[a]	Aglycone nucleus	Sugar residue[b]
Betanin group	Betanidin	——
Betanin group	Betanin Phyllocactin Lampranthin I	5-*O*-Glc 5-*O*-Glc 5-*O*-Glc
Amaranthin group	Amaranthin Celosianin II	5-*O*-Glc-2-*O*-GlcU 5-*O*-Glc-2-*O*-GlcU
Bougainvillein	Bougainvillein	5-*O*-Glc
Gomphrenin group	Gomphrenin I	6-*O*-Glc
Betaxanthins	Dopaxanthin Indicaxanthin Portulaxanthin II Vulgaxanthin I	DOPA Proline Glycine Glutamic acid

[a]Names were standardized by Strack et al
[b]*Abbreviations*: *Glc* β-D-glucose, *GlcU* β-D-glucuronic acid, *DOPA* 3,4-dihidroxyphenylalanine

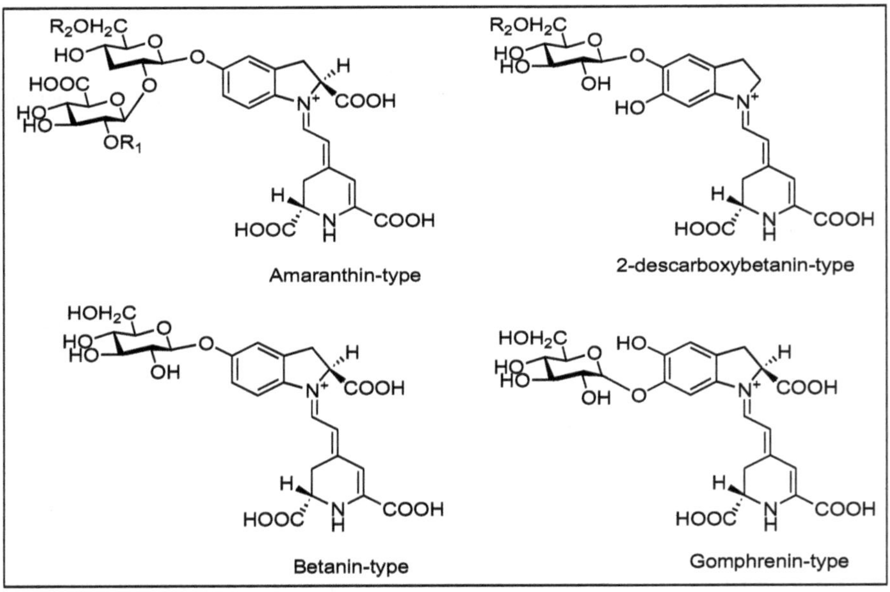

Fig. 1.7 Four groups of betacyanins

Chromophore by the Chemical Structure

By considering the chemical structure of chromophore pigments, they are classified as:

- Conjugated system of chromophore: Anthocyanins, caramel, and carotenoids
- Lakes and synthetic pigments
- Porphyrins coordinating with metals: Chlorophyll, myoglobin, and derivative of them.

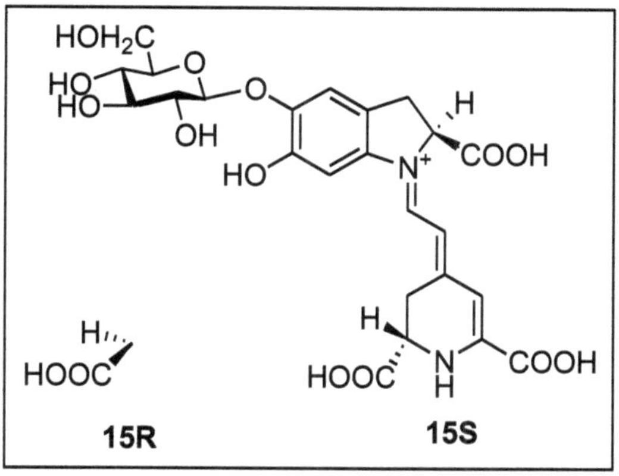

Fig. 1.8 Chemical structure of betanin (15S; isobetanin 15R)

Table 1.4 Examples of betalains

Entry	Name	Substituent group[a]		Botanical source
		R_1	R_2	
Betacyanins	Betanin	β-Glucose	H	*Beta vulgaris*
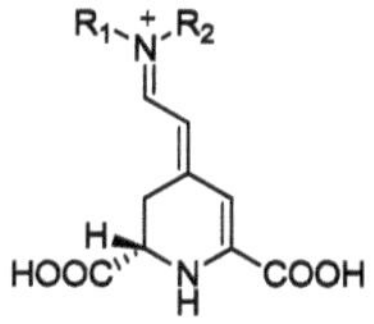	Phyllocactin	6-*O*-(Malonyl)-β-glucose	H	*Phyllocactus hybridus*
	Lampranthin I	6-*O*-*p*-Coumaroyl-β-glucose	H	*Lampranthus* spp.
	Amaranthin	2-*O*-(β-glucuronic acid)-β-glucose	H	*Amaranthus tricolor*
	Celosianin II	2-*O*-[*O*-(trans-feruloyl)-β-glucuronic acid]-β-glucose	H	*Celosia cristata* L.
Betaxanthins	Indicaxanthin	Both groups together form proline		*Opuntia ficus-indica*
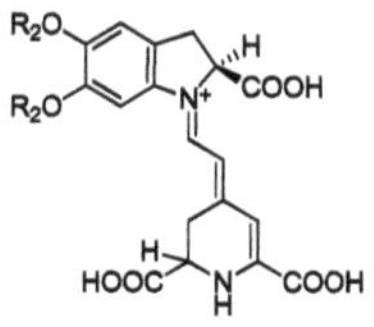	Portulacaxanthin I	Both groups together form hydroxyproline		Portulaca grandiflora
	Vulgaxanthin I	H	Glutamine	*Beta vulgaris*
	Vulgaxanthin II	H	Glutamic acid	*Beta vulgaris*
	Dopaxanthin	H	L-DOPA	*Glottiphyllum longum*

[a]R_3 and R_4 may be acyl or glycosyl substituent groups

Structural Characteristics of the Natural Pigments

By taking into account structural characteristics, natural pigments can be classified as derivatives of:

- Tetrapyrrole: Hemecolors and chlorophylls
- Isoprenoid: Iridoids and carotenoids

N-Heterocyclic compounds other than tetrapyrroles: betalains, flavins, phenoxazines, pterins, phenazines, and purines. Oxygenated heterocyclic compounds (like benzopyran derivatives): flavonoid pigments, i.e., anthocyanins and other quinones, anthraquinone, benzoquinone, naphthoquinone, and melanins [16].

Food Additives
By considering the pigments as food additives, their classification by the FDA is certified: they contain two subgroups which are man-made that include synthetic pigments and lakes.

Relieved from Certification
This group includes pigments derived from natural sources such as vegetables, minerals, or animals and man-made counterparts of natural derivatives [3].

References

1. Khan, M. I., & Giridhar, P. (2015). Plant betalains: Chemistry and biochemistry. *Phytochemistry, 117*, 267–295.
2. Gandía-Herrero, F., Escribano, J., & García-Carmona, F. (2016). Biological activities of plant pigments betalains. *Critical Reviews in Food Science and Nutrition, 56*(6), 937–945.
3. Delgado-Vargas, F., Jiménez, A. R., & Paredes-López, O. (2000). Natural pigments: Carotenoids, anthocyanins, and betalains—Characteristics, biosynthesis, processing, and stability. *Critical Reviews in Food Science and Nutrition, 40*(3), 173–289.
4. Hatlestad, G. J., & Lloyd, A. (2015). *Pigments in fruits and vegetables. Genomics and dietetics*. Network: Springer Chapter 6 The Betalain Secondary Metabolic pages.
5. Jain, G., & Gould, K. S. (2015). Are betalain pigments the functional homologues of anthocyanins in plants? *Environmental and Experimental Botany, 119*, 48–53.
6. Strack, D., Vogt, T., & Schliemann, W. (2003). Recent advances in betalain research. *Phytochemistry, 62*, 247–269.
7. Esatbeyoglu, T., Wagner, A. E., Schini-Kerth, V. B., & Rimbach, G. (2015). Betanin—A food colorant with biological activity. *Molecular Nutrition & Food Research, 59*(1), 36–47.
8. Herbach, K. M., Stintzing, F. C., & Carle, R. (2006). Betalain stability and degradation—Structural and chromatic aspects. *Journal of Food Science, 71*(4), R41–R50.
9. Hughes, N. M., & Lev-Yadun, S. (2015). Red/purple leaf margin coloration: Potential ecological and physiological functions. *Environmental and Experimental Botany, 119*, 27–39.
10. Davies, K., Zryd, J. P., & Christinet, L. (2004). *Plant pigments and their manipulation, Annual Plant Reviews, 14*. Oxford: Blackwell Publishing Chapter 6 Betalains pages.
11. Pavokovi, D., & Rasol, M. K. (2011). Complex biochemistry and biotechnological production of betalains. *Food Technology and Biotechnology, 49*(2), 145–155.
12. Tanaka, Y., Sasaki, N., & Ohmiya, A. (2008). Biosynthesis of plant pigments: Anthocyanins, betalains and carotenoids. *The Plant Journal, 54*, 733–749.
13. Gandía-Herrero, F., & García-Carmona, F. (2013). Biosynthesis of betalains: Yellow and violet plant pigments. *Trends in Plant Science, 18*(6), 334–343.
14. Gengatharan, A., Dykes, G. A., & Sim-Choo, W. (2015). Betalains: Natural plant pigments with potential application in functional foods. *Food Science and Technology, 64*, 645–649.
15. Delgado-Vargas, F., & Paredes-López, O. (2002). *Natural colorants for food and nutraceutical uses*. Boca Raton: CRC press.
16. Clement, J. S., & Mabry, T. J. (1996). Pigment evolution in the Caryophyllales: A systematic overview. *Plant Biology, 109*(5), 360–367.

Chapter 2
Sources of Betalains

2.1 Occurrence

Betalains occur only in few plant families of order Caryophyllales where they found vital constituents in edible parts of plants usually in leaves, flowers, and stems. They also occur in some fungi of higher order like in the fly agaric (*A. muscaria*) and some genera of *Basidiomycetes*. Red beetroots (*B. vulgaris*), cacti fruits belonging to the genus *Opuntia* (mainly *Opuntia ficus-indica*), the dragon fruits of *Hylocereus* cacti (mainly *Hylocereus polyrhizus*), and the Swiss chard (*B. vulgaris*) are the known edible sources of betacyanins and betaxanthin. The less common edible sources are Ulluco tubers (*Ullucus tuberosus*), fruits and berries of *Eulychnia* cacti, and *Rivina humilis*. It is a class of yellow and red indole-derived pigments which replace anthocyanin pigments in plants and are mainly responsible for coloration, but their role in fungi is unknown. Betalain accumulation occurs in cell vacuoles synthesized mainly in epidermal and subepidermal plant tissues due to their hydrophilicity.

2.2 Aizoaceae

Aizoaceae is also called *Ficoidaceae*, the ice plant family or fig-marigold family. It belongs to Magnoliopsida flowering plants family. It contains 1900 species and 135 genera. Other names of this family are stone plants, carpetweeds, or vygies. Due to their resemblance with stones or pebbles, they are sometimes called mesembs. Due to their glistening nature, several species are also recognized as "ice plants" [1]. Aizoaceae family consists of small shrubs or juicy herbs. A number of stems have basal origin of the plant. Mat-forming stems are also observed a few times. Vascular bundles are present in the form of concentric rings. They have either alternate or opposite leaves. They possess either equal pair of leaves (*Sesuvium*) or unequal pair of leaves (*Trianthema*) [1]. It consists of the following major genera as shown in Fig. 2.1.

E. Akbar Hussain et al., *Betalains: Biomolecular Aspects*,
https://doi.org/10.1007/978-3-319-95624-4_2

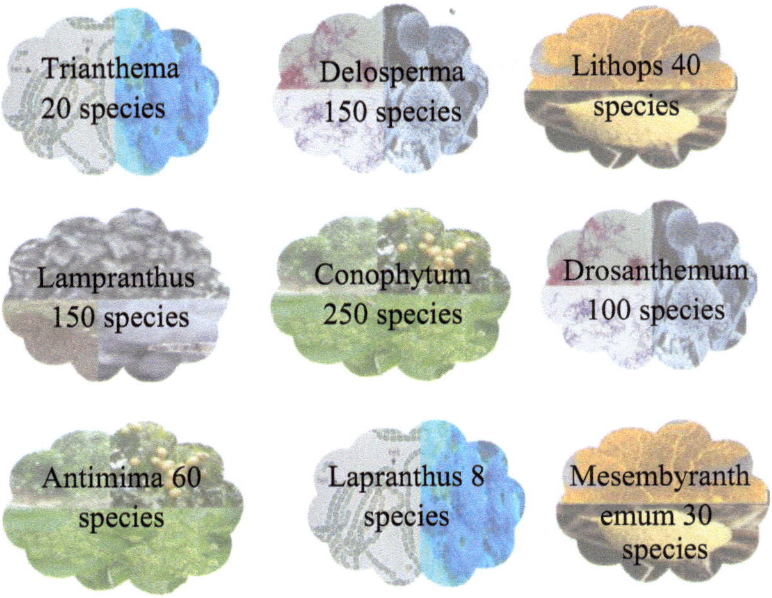

Fig. 2.1 Major genera of Aizoaceae showing the number of species

Inflorescence of this family is in pair or solitary or terminal or axillary. Flowers are usually bisexual. Calyx consists of five connate sepals often imbricated with thorny protuberance on black-blue tip. Corolla is usually absent which is usually represented by petaloid and staminodes. Androecium usually consists of five to numerous stamens which arise from hypanthium. Pollen is tricolpate. Gynoecium comprises of two to five joint carpels. Ovary may be inferior or superior. Placenta may be basal or parietal, having 2–5-locular, and axile. Ovules range from one to many. It is inverted to campylotropous. Usually, there exists a disc.

The fruit of this family is loculicidal, capsulated, circumscissile, or septicidal. Occasionally the fruit is berry and a nut but not often. Seed is usually present with great embryo having a curve shape. Endospermal absence is observed which is replaced by perisperm. The phenomenon of pollination is usually occurred by bees, wasps, butterflies, and beetles. Wind or water is the source for the dispersal of seed.

Monophyletic linkage is usually observed in this family which is represented in the form of distinct clades. These classes have been recognized in subfamilies of Aizooideae, Mesembryanthemoideae, Sesuvioideae, and Ruschioideae.

Members of this family which possess numerous petaloids and staminodes are located in Ruschioideae and Mesembryanthemoideae which make a monophyletic group. Those members which have circumscissile capsules and arillate seeds are placed in *Sesuvium* and linked genera (Sesuvioideae) which constitute a clade. Genus *Mesembryanthemum* contains more than 100 species, but it has been further split into 532 plants; according to systematic numerous genera, a bulk of genera are placed in Ruschioideae [1].

The chemical constituents which have been identified are betacyanins and hydroxycinnamic acid derivatives [3]. Betalain-isolated structures are hydroxycinnamic acids and aliphatic acid (malonic acid, 3-hydroxy-3-methylglutaric acid, and citric acid). Sometimes they are also present as acyl constituents. It has been reported that four novel acylated betacyanins are isolated from flowers of *Drosanthemum floribundum* [2].

Zaleya decandra L. (Aizoaceae) is a prostrate mainly found in world's subtropical and tropical areas and in South India. Orchitis, asthma, and hepatitis are treated by using the root portion of this plant. It is also used to relieve partial headache by dropping the juice of leaves into nostrils. Its callus culture had been well-thought-out as one of the potential betalains' fabricator which shows high radical scavenging activity. We can obtain desirable medicinal compounds from plants by tissue culture technique. Medicinal plants are the major source of crude drugs and extracted on a global scale. Many active and potent compounds are isolated from plants which usually include many alkaloids like pain killer morphine, antitussive codeine, phosphodiesterase inhibitor papaverine, and several kinds of cardiac glycosides as cardiac insufficiency [3].

Some species of Aizoaceae family are used for ornamental purposes such as *Mesembryanthemum* (ice plant), *Lampranthus*, *Dorotheanthus*, and *Carpobrotus*. *Tetragonia* is used as vegetable. Some species play an important role in the stabilization of sand dunes and road banks [1].

2.3 Amaranthaceae

Amaranthaceae belongs to evergreen, perennial herbs. An interesting feature of this family is that they have grown for edging beds and for bedding, for example, *Alternanthera* [4]. It has been found that several important genera are present in Amaranthaceae family as shown in Fig. 2.2.

The members of this family have been mainly distributed in subtropical and tropical areas of South America and Australia. The family Amaranth*aceae* consists of 72 genera and 1020 species. It has been mainly distributed in tropical areas, which center in Africa and America. *Amaranthaceae* consists of small herbs or small shrubs which contain swollen nodes and unilacunar nodes. Xylem and phloem are

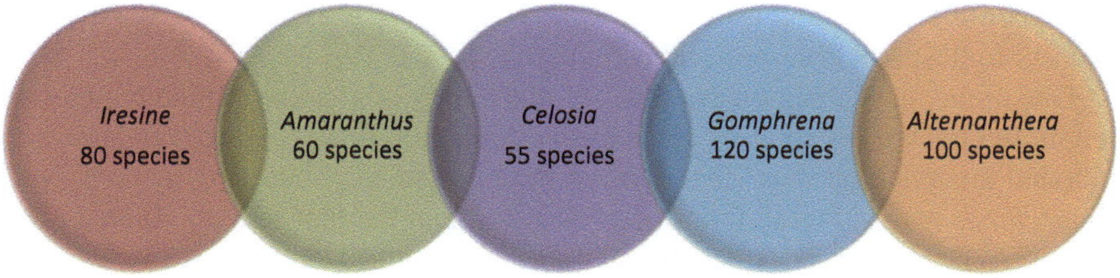

Fig. 2.2 Important genera of family *Amaranthaceae* showing the number of species

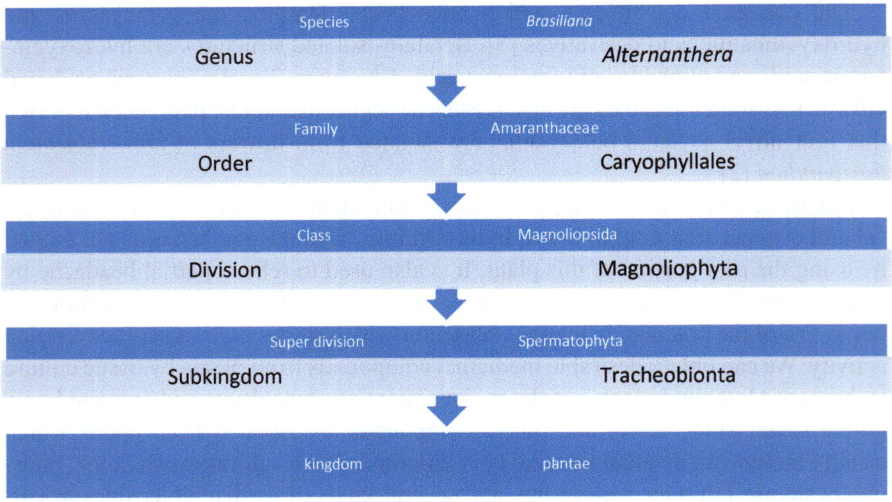

Fig. 2.3 Taxonomic classification

present in concentric ring arrangements. Xylem and phloem are known as vascular bundles which are usually present and contain sieve-tube plastids PIII-A type. It has been observed that it contains betalains instead of anthocyanins [9] (Fig. 2.3).

They have alternate or opposite, herbaceous, petiolate to sessile, simple, entire leaves. Stipules are absent. The family has cymose inflorescence. The presence of spikes or panicles has also been observed. Flowers are small, greenish, and bisexual, and rarely does it happen that they are unisexual. They have actinomorphic symmetry and are hypogynous [9]. Sepals and petals are absent. Perianth consists of three to five free or joint tepals which are typically lasting, and occasionally *Ptilotus* (accrescent) in fruit is generally scarious and dry. Androecium contains five stamens; rarely it ranges from three or even six to ten stamens which are opposite the tepals. It has been observed that filaments are slightly basally connected, sometimes adnate to tepals and anthers, and are inflexed in bud, bithecous (*Amaranthus*) or monothecous (*Gomphrena*), with longitudinal dehiscence. Pollen grains are multiporate. Often one to three numbers of staminodes and spinulose are present. Two to three joint carpels are present in gynoecium which contains unilocular, superior ovary usually comprising of single ovule. It has basal placentation. It has been observed that it consists of styles ranging from one to three. The fruit of this family is a nut or utricle or circumscissile capsule. The embryo is usually curved and spiral shaped. Endosperm is absent, and perisperm is present [9].

One of the most common examples of this family is amaranth which is a dicotyledon. In Greek translation its name means "unfading flower" which contains quinoa and amaranth species. Taxonomists have classified it into two sections, viz., *Amaranthus* and *Blitopsis*. It has been observed that the first section, *Amaranthus*, includes grain types in which inflorescence is terminal and the second section *Blitopsis* has leaf types and flowers which are borne in clusters in leaf axils. Many parts of the plants are excellent sources of compounds like vitamins A and C, calcium, and iron.

It has been mainly distributed in the Mexico, China, Russia, Peru, Guatemala, Kenya, and India. These are very advantageous to humans, and the examples are *A. caudatus* L., *A. cruentus* L., and *Amaranthus hypochondriacus* L. *A. hypochondriacus* is one of the most dominant species found in Mexico. Mainly, it is present in warm areas of Puebla, Guerrero, and Morelos states. Due to its temperature, drought, and saline solid resistivity, this plant has many agrarian benefits [10].

Another most important example of this family is *Alternanthera brasiliana* recognized as an herb and is found mainly in Brazil. It is prostrate and perennial and constitutes a rounded to quadrilateral stem, swollen nodes, and long internodes having opposite leaf attachments. Its inflorescence is cymose that has been observed as hermaphrodite, actinomorphic, and monocyclic flowers [4].

In Greek, plants of *Amaranthus* or amaranths are meant as "never-fading flowers" [6]. A number of amaranth species are often present in the form of weeds, while in most parts of India, amaranth is used as leafy vegetable.

Amaranthaceae family is meticulously related to Chenopodiaceae (which is positioned in APG classifications to be similar to Amaranthaceae); nonetheless it is distinguished by means of scarious perianth and bracts. The presence of staminodes and connate stamens has been observed. Many taxonomists gave different perceptions about this family. One of them is Hutchinson, who believed that it has originated from caryophyllaceous family tree, and another one is Cuenoud, who believed that Amaranthaceae is monophyletic which is reinforced by chloroplast DNA restriction sites, morphology, as well as *rbcL* sequences.

Evidence is the resemblance of the pollen with stiffened tecta, a thickened endexine and openings with reduced sharp specks of exine underlain by lamellar plates. It has been observed that the parting of Amaranthaceae and Chenopodiaceae is random. Some researchers like Pratt and coworkers ponder Amaranthaceae as polyphyletic. According to Thorne, Amaranthaceae is distinct from Chenopodiaceae as three subfamilies are included in it: Gomphrenoideae, Amaranthoideae, and Polycnemoideae [9].

Chemical constituents which have been identified are betaxanthin and methyl derivative of arginine betaxanthin (yellow), amaranthin (red-violet), carotenoids, and anthocyanins in red *Amaranthus* (A. tricolor) [8, 9].

Other constituents which have been identified are minerals, amino acids, alkaloids, saponins, carotenoids, tannins, phytosterols, waxes, terpene alcohols, and carbohydrates such as starch, sugars, and lipids. It also includes phospholipids, triacylglycerols, squalene, phenolic substances, and lipid-soluble vitamins, for example, tocopherols that contribute a major role in the lipophilic region which are thought to be present in roots [10].

Betalain-isolated structures are 3 betaxanthins and 16 betacyanins (isoamaranthine or amaranthine) from *Amaranthus tricolor* (vegetable amaranth). It has been observed that betacyanins consist of two betanin types, six gomphrenin types, and eight amaranthine types, and it has been further distributed into ten acylated and six non-acylated (simple) betacyanins [6].

Many members of this family have been used for dietary purposes as grains (*Amaranthus*) and vegetables (*Boerhavia*, *Amaranthus*) and also a source of animal fodder [9]. One of the most important examples of this family is red amaranth

(*Amaranthus tricolor*) which is nutritive and has high minerals such as vitamin C, calcium, and iron and pigments such as flavonoids, betalains, and carotenoids.

It has been observed that another most important example is green amaranth (*Amaranthus viridis*) which has been used as an anti-allergic, anti-ulcerogenic, anti-inflammatory, hepatoprotective, and antiviral and some amount of minerals, vitamin C, carotenoids, and phenolic acids are also present [7]. Many members of this family are source of medicinal plants such as *Amaranthus spinosus* (*A. spinosus*) [5].

Many members of this family have been used for ailment of jaundice, hepatic disorders, malaria, and scanty urine or for therapy of injuries. Many members of this family show high antioxidant activity observed in Amaranthus betacyanins in modeling food system. High antioxidant activity has been showed by almost 19 diverse kinds of betalains from *Amaranthaceae* family [8].

It has been observed that vegetables and fruits used as human diet contain natural antioxidants which provide immunity and protection against different disorders such as Alzheimer's disease, diabetes, cardiovascular disease, cancer, age-related functional decline, dysentery, cataracts, diarrhea, night blindness, hazy vision, and postnatal complaints.

A number of plant parts which are a source of various substances such as green leafy vegetables can be used as a basis of vitamins, minerals, and antioxidants. It has been observed that *Amaranthus viridis* (green) and *Amaranthus tricolor* (red) have been used as flavoring agent and in many pharmacological activities [7]. The taste of cooked or raw amaranths resembles with the spinach or other leafy vegetables. It has been observed that the plant can be used as a febrifuge, abortifacient, cholagogue, and galactagogue.

One of the most important uses is in indigestion. Many leaves of the plants are good sources of fodder due to which milk production in cattle increases [4]. One of the important roles of this family is to increase the activity of sex hormones. Many members of this family play an important role in cosmetics such as carotenoids, an important constituent of beetroot, which enhance the quality, growth, thickness, and shine of the hair. It has been observed that beetroot enhances the skin fairness and flawlessness [6].

It has been noted that many members of this family also play an important role in photosynthesis such as *Amaranthus*, an annual plant with C4 type of photosynthesis which is responsible for high production potential and makes photosynthesis fast, and an interesting feature is that it can grow in adverse conditions. Many members of this family can be used in cosmetology and biodegradable plastics. Lectin which is an important chemical constituent of this family has been used frequently in cell biology, cancer research, and immunology due to its biological activities, for example, cytotoxicity, mitogenicity, and immunosuppression.

Another most important component is *A. hypochondriacus* lectin which has negative mitogenic activity observed in *A. caudatus* in hog lymphocytes, while positive activity is observed in mouse spleen lymphocytes. The presence of tocopherols and tocotrienols has been observed in the seed which showed antioxidant activity (α-tocopherol), antitumor activity, cholesterol and synthesis suppression; regulatory levels of serum cholesterol decrease the low-density lipoprotein-cholesterol production and tend to regulate the lipoprotein lipase enzyme [10].

Many species have been investigated and showed that they can be used for ornamentals purposes which include *Gomphrena* (globe amaranth), *Amaranthus* (amaranth), *Iresine* (bloodleaf), and *Celosia* (cockscomb), and another one most important example is the species of *Tilanthera* and *Alternanthera* which have been developed as edge plants consisting of ornamental leaves [9].

2.4 Basellaceae

It's a family containing flowering plants of the order Caryophyllales in the core eudicot clade, conferring to the phylogenetic assembly of angiosperms. In this family 4 genera are present with 19 different herb species, some having mounting habits as shown in Fig. 2.4 [12].

Malabar spinach and ulluco are economically important of this family. In this plant some chemicals such as kaempferol (flavonoid), betalains, and basellasaponins are present. A betalain pigment gomphrenin derivative is present in Basella fruit [12]. In leaf, vitamins A, C, E, K, and B9 (folic acid), proteins, fat, riboflavin, thiamine, niacin, and minerals, i.e., phosphorus, magnesium, calcium, and iron, are present. In Basella, kaempferol flavonoid; basellasaponins; amino acid, i.e., lysine, arginine, isoleucine, leucine, tryptophan, and threonine; phenolic compounds; and peptide are present [13]. The seeds are found to be rich in oleic acid. The fatty oil from seeds was found to contain palmitic, oleic, linoleic, and linolenic acid. The proteins revealed presence of lysine, threonine, valine, methionine, and leucine [15]. Mixture of starch-type glucan and polysaccharides is present in mucilage of *B. alba* which can be detached by complex of starch iodine. Mucilage of plant is composed of polysaccharides that are soluble in water [12].

Gomphrenin (Fig. 2.5) is one of the derivatives of betalain [18].

Fig. 2.4 Four genera of family Basellaceae

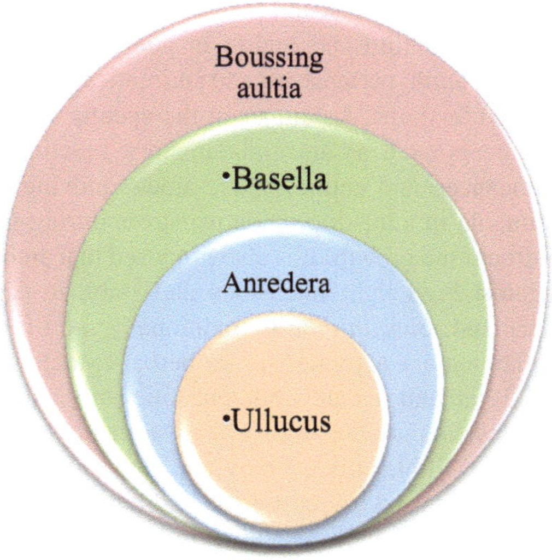

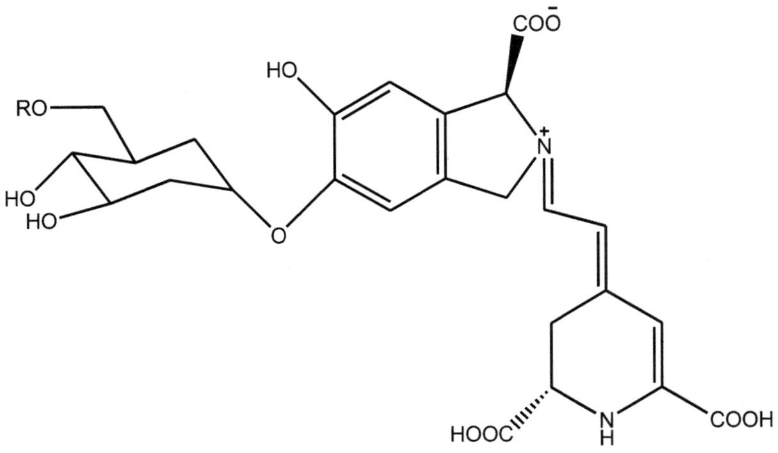

Fig. 2.5 Gomphrenin

A number of chemical compounds obtained from these species, like water, ethanol, chloroform, and petroleum, are used for various pharmacological actions. One important medicinal herb is *Basella alba*, which is used for therapeutic purposes or as a precursor for synthesis of various drugs like other medicinal plants due to the presence of many pharmacologically important substances, for example, proteins, carbohydrates, oils, fats, enzymes, vitamins, minerals, carotenoids, flavonoids, alkaloids, terpenoids, quinones, sterols, saponins, tannins, simple phenolic glycosides, and polyphenols. The mucilage is used as a water retention agent, gelling agent, thickener, film former, and suspending agent for medicinal purpose. It also acts as anti-freckle antioxidant for cosmetic development and acts as anti-inflammatory agent as contemporary treatment. β carotene is also present in *B. alba*. The fiber matter of these plants is responsible for bulk in food and thus supports to decrease the consumption of starchy diet, prevents constipation, reinforces digestive purpose, and hence lessens the occurrence of metabolic ailments like hypercholesterolemia and diabetes mellitus beginning in adulthood. These are also potent blood-building agents, antihypertensives, and antibiotics and progress fecundity in ladies when consumed in soups. A number of therapeutic benefits are associated with anthocyanin pigments such as anti-inflammatory function, vaso-protective, chemo-protective, anticancer, and antineoplastic assets, moving back stage correlated shortages and valuable in adjusting oxidative stress during gestation problems by retardation of intrauterine growth. It is also proposed that anthocyanins can protect the chloroplast against high light intensities and stabilize triple-helical complexes of DNA. In Thailand, *Basella alba* is traditionally used for antioxidant, cytotoxicity, and anti-inflammatory activities. From herbal extracts, it is considered that obtained flavonoids are antioxidant while phenolic compounds are antimutagenic agents [11].

Basella mucilage is an effective medicine for bruise, irritant, laboring, and ringworm. Mild laxative, antipyretic, and diuretic purposes are fulfilled by leaves and stem. It is used as antipruritic and burn treatment in India and as freckle and acne treatment in Bangladesh [12]. In India, leaves and stem of *B. alba* are used as anticancer, i.e., leukemia, oral cancer, and melanoma [14].

The pharmacological property of *B. alba* is that it is an excellent source of calcium, magnesium, iron, and vitamins A, C, and B9 (folic acid) [16]. It also possesses good antimicrobial activities against various pathogenic organisms [17].

2.5 Cactaceae

The family Cactaceae includes mostly spiny succulents and stems having photosynthetic ability, consisting of 1000–2000 species and 30–200 genera which supplementary categorized *p*-plastid and betalain presence. In general, three subfamilies arise from Cactaceae, Opuntioideae, Pereskioideae, and Cactoideae, while a fourth group has been added recently, the subfamily Maihuenioideae. Its distribution ranges from the south of the Arctic Circle in Canada to the slope of Patagonia in South America and nearly in every type of habitat in between these, mostly in xeric areas. Some epiphytic species such as pencil cactus (Rhipsalis) that are exceptional are found in Africa, Madagascar, and Ceylon; they grow at the altitudes, above sea level, such as the Valley of Death to over 4800 m in the Andes. Their habitat is at the climates where there is annual rainfall perception more than 500 cm or even no measurable rainfall. Typically spiny plants are founded at arid and moist regions and spineless with some epiphytes in tropical areas.

Hylocereus spp. (pitayas) is a climbing, perennial cactus plant living in Central, South, and North America's tropical regions. It is important as both food crop and ornamental plant. Its fruits are a good source of natural pigments and are also important in food processing due to high betalain contents present in them. Another example of this family is *Hylocereus polyrhizus* commonly known as red pitaya or dragon fruit. It is from the subfamily Cactoideae having red-purple-colored flesh and black seeds [24]. Dragon fruit has obtained attention mostly in Asian countries, due to its color, nutritional values, and other features. Its most vital role is as an antioxidant [25].

Water-soluble, red-violet betacyanin pigments are present in the fruits and flowers of many species of the Cactaceae family. Structurally, these are betalamic acid immonium conjugates with cyclo-DOPA or frequently O-glucosylated cyclo-DOPA. Glucosylation at C5 position is a characteristic for betanin-type betacyanins named after betanin, the simplest betacyanin. Other types are gomphrenin-type betacyanins, which are 6O-glucosylated derivatives and amaranthine-type derivatives [19, 20]. Other studies also indicated the presence of indicaxanthins, carotenoids, phenolic contents [26], minerals, and several types of amino acids (alanine, arginine, and asparagine). Important vitamins include vitamins C (ascorbic acid), E, and K and beta-carotenes [29].

Betacyanin (Fig. 2.6) accumulates during flower development as the malonyl derivatives and phyllocactin and 2′-apiosyl-phyllocactin as the main pigments in the petals. The dark red carpels contained nearly double the amount of total betacyanins and traces of less polar feruloyl betacyanins. Feruloyl betacyanins are the predominant components in the extract from stamens.

Pitaya contains betalain pigments. The red-violet betacyanins belong to betalain pigments. The red color of pitaya is attributed to betacyanins [23].

Fig. 2.6 Betacyanin

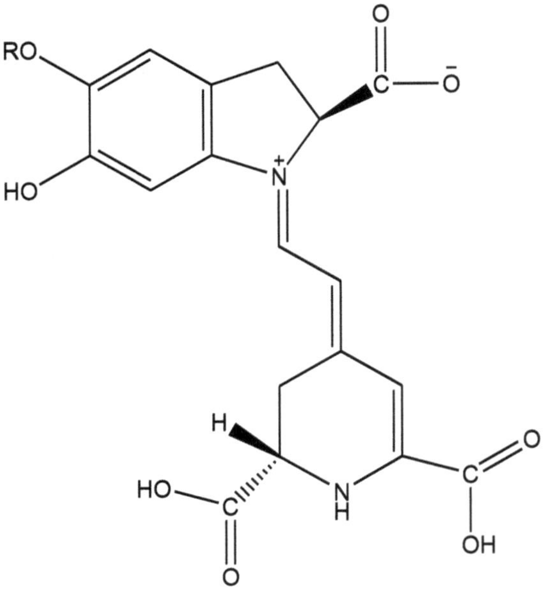

Curing certain disorders, such as calciuria, to prevent osteoporosis and improve bone mineral density, preventing food contamination by *V. cholerae* and Campylobacter and treating gut tract disorders associated with these microorganisms by destroying their cell membranes can be done via pharmacological profile of *O. ficus-indica.* It is also used as a traditional medicine for curing wounds, burns, hyperlipidemia, edema, catarrhal gastritis, and obesity. Hypoglycemic, antiviral, and anti-inflammatory activities are associated with alcoholic extracts [21].

Regular consumption of dragon fruit helps in fighting against cough and asthma. Also it helps for healing wounds and cuts quickly due to high amount of vitamin C in it. It has an important role to enhance immune system and to stimulate the activity of other antioxidants in the body. Moreover, dragon fruit is also rich in flavonoids that act against cardio-related diseases; also dragon fruit aids to treat bleeding problems of vaginal discharge. Dragon fruit is also packed with B vitamin group (B1, B2, and B3) which possesses an important role in health benefit.

Increment of carbohydrate metabolism and energy production is associated with vitamin B1, whereas vitamin B2 is present as multivitamin in dragon fruit, and it helps to recover and increase appetite loss. Vitamin B3 present in dragon fruit plays an important role in lowering bad cholesterol levels; it provides and moisturizes smooth skin appearance as well as improves eyesight and prevents hypertension.

Type 2 diabetes patients are given dragon fruit to reduce the levels of blood sugar; studies suggest that the glucose found in this fruit helps in controlling the blood sugar level for diabetes patients. Dragon fruit contains high levels of phosphorus and calcium, which helps to reinforce bones and plays an important role in tissue formation and forms healthy teeth [22].

Antiproliferative and antioxidant activities are associated with *H. monacanthus* fruits. Its pulp is a source for phytochemically bioactive compounds (antioxidants), while the peel helps to retard growth of cancer-affected cells. Prebiotic properties, partial resistance to human α-amylase and reduction of stomach acidity, are associated with oligosaccharides of *H. monacanthus* and *H. undatus*, and they also promote Lactobacillus and Bifidobacteria [25].

In pharmaceutical, cosmetic, and food industries, *Opuntias* are used [27]. In addition, antiatherosclerotic, antiulcerogenic, hepatoprotective, immunomodulatory, and hypoglycemic activities are also associated with it [28].

2.6 Caryophyllaceae

Caryophyllaceae includes in the dicotyledon order Caryophyllales in the APG III system. It is commonly known as carnation or pink family that is comprised of outsized family of mostly flowering herbaceous plants comprising about 2625 known species and 81 genera. It is present mainly in temperate regions, while some species can also occupy tropical peaks. Campions (Lychnis and Silene) and fire pink and pinks and carnations (Dianthus) are some of the most generally recognized participants of this family. Some of its species are widespread wildflowers; in general these are ornamental plants. A number of its species are developed in bordering and Mediterraneanregions of Asia and Europe. In the Southern Hemisphere, a small number of species and genera are present, although the Antarctic pearlwort (Colobanthus quitensis) is included in this family, that is, the biosphere's southernmost dicot, and is one of only two floral plants present in Antarctica [1]. Normally these are herbs having opposite leaves; swollen nodes; dichasial cyme inflorescence; ten or lesser stamens; corolla caryophyllaceous, unilocular, superior ovary with free central placentation; obdiplostemony; and fruit with an opening of capsule by the teeth or valves. The major genera of this family are shown in Fig. 2.7 [1].

(Chenopodiaceae, Achatocarpineae, Caryophyllineae, Phytolaccaceae, and Cactineae) within Caryophyllaceae and Caryophyllales that are kept in monotypic Caryophyllineae suborder [1].

2.7 Chenopodiaceae

Chenopodiaceae is one of the examples of family of flowering plants. Other name of this family is goosefoot family. Among most plant classifications, it is widely recognized (notably the Cronquist system). These plants have been placed in the family Amaranthaceae according to the advanced APG system (1998) that is based on gene and APG II system (2003). According to molecular phylogeny, Chenopodiaceae is paraphyletic.

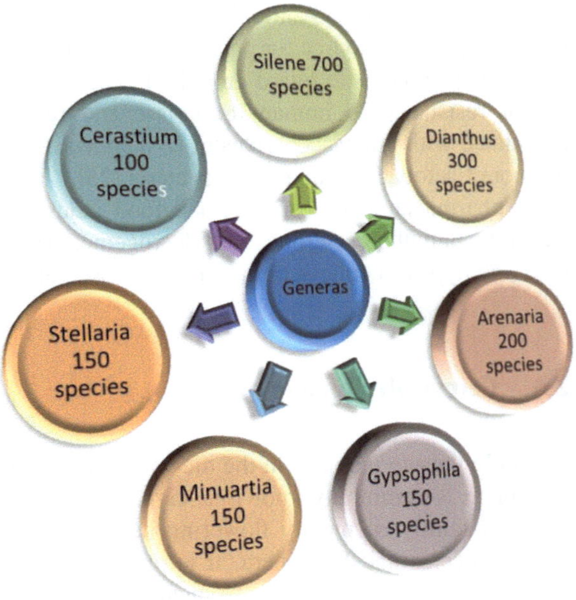

Fig. 2.7 Genera of family Caryophyllaceae showing the number of species

The members of this family have been mainly found in tropical and temperate regions; mostly these are present in arid and semiarid briny locales. In the Andean region of South America, quinoa (Chenopodium quinoa Willd.) plants have a large variation in genes which can grow under very severe and drastic ecological circumstances such as hail, frost, drought, and high loftiness [1].

Chenopodiaceae family consists of 97 genera and 1305 species. This family consists of the following major genera: Atriplex consists of 300 species, Salsola consists of 120 species, Chenopodium consists of 105 species, Suaeda consists of 100 species, and Salicornia consists of 35 species.

Chenopodiaceae family comprises of small herbs or shrubs but rarely consists of Haloxylon (small trees) which is typically distributed in salty habitations. Occasionally, most of the plants are succulents (Salicornia). Their nodes are usually unilocular, and these are frequently roofed with whitish bloom. Bundles of xylem and phloem are present in the form of concentric rings. PIII-C-type sieve tube plastids usually contain betalain instead of anthocyanins. They have cuticle waxes with platelets which provide protection.

They have alternate, minute to large, and rarely opposite leaves (Salicornia, Nitrophila). Stipules are absent. They are either petiolate or sessile. Sometimes either fleshy or scaly leaves are present. Stipules are usually absent [1]. Cymose inflorescence is the characteristic of this family. Flowers are green, small, and mostly bisexual but rarely are unisexual. Plants are actinomorphic, dioecious or monoecious, and hypogynous. Sepals and petals are absent, but two to five united tepals are present. It rarely happens that tepals are free (Salsola), not united. They are herbaceous and persistent. Sometimes either appendage with tubercles, spines, or wings is either present or absent.

Androecium consists of five stamens. It is very rare that three are opposite to lobes of perianth. Usually the filaments are free. Gynoecium mostly consists of two carpels which are seldomly joint up to five carpels. Unilocular, superior ovary is present containing a single ovule. Basal placentation is characteristic of this. Styles range from two to five. The fruit of this family is mostly utricle or nut when encircled in membranous perianth. Embryo is usually spiral-shaped. Endosperm is absent and perisperm is present [1].

Chemical constituents which have been identified are free and bound phenolics and flavonols which are extracted from their colored quinoa samples [30]. The other chemical constituents are betaxanthin which has composition of red and yellow beetroots which are yellow in color. Other families also contain lysine-bx, methionine-bx, and aspartic acid-bx (miraxanthin II) till now [31].

Quinoa seeds have great biological importance because of its highly nutritious grain quality all over the world and high quality of protein as we know that essential amino acids are very important and are the main building blocks. Different nutritional compositions and bioactive compounds have been observed among different ecotypes.

Due to high contents of phenolic material, members of this family show antioxidant capacity which varies according to environmental conditions and variations in genes. In quinoa seeds, phenolic antioxidants are existing in free or in bound state when attached with the structure of cell wall [30].

Many members of this family are source of plant food. One such example is *Beta vulgaris* which is used as leafy vegetable. Many members have been observed as a good wormseed source which acts as a lamb's-quarters such as *Chenopodium album* (bathua in Hindi) and vermifuge such as *Spinacia oleracea*. Several leaves and seeds of *C. quinoa* are edible [1].

2.8 Nyctaginaceae

The family Nyctaginaceae is also called the 4-o'clock family which belongs to the family of flowering plant. An attractive feature of this family is that it has "anthocarp," that is, a unique type of fruit. It has been observed that large (>1 00 μm) pollen grains are produced in many genera [1].

As family, it has been widely recognized by many taxonomists. Recently it has been placed in order Caryophyllales in core eudicot clade following the APG II system (2003). Many genera of this family have some common characteristics such as sticky bands between the nodes on the stems. One of the most interesting features is the cleistogamous flowers; they are self-pollinated without opening. They possess potential to grow on soils with elevated level of gypsum [1].

It consists of 400 species and 31 genera which are distributed in subtropical and tropical areas, especially in the New World, but some are present in temperate regions. Herbaceous genera are mainly found in arid North America which is studied and confirmed by Douglas and his colleague who explained and demonstrated the relationship among whole genera of this family and showed their diversification [1].

Its major genera are *Neea* consisting of 80 species, *Guapira* consisting of 70 species, *Mirabilis* consisting of 45 species, *Pisonia* consisting of 40 species, *Abronia* consisting of 30 species, *Boerhavia* consisting of 20 species, and *Bougainvillea consisting of 80 species.*

This family mostly consists of herbs (*Boerhavia*, *Mirabilis*) having swollen nodes which is the most attractive feature of this family. It also consists of shrubs, for example, woody climbers (*Bougainvillea*) or *Pisonia* and rarely trees such as *Pisonia alba* are also present.

Vascular bundles are present in the form of concentric rings constituting raphide crystals of calcium oxalate and betalains. One of the interesting features is woody oxidizing which means when cut, it turn orange to reddish-brown [1].

Roots are branched and may be taproot which may be fusiform (*Boerhavia*) or tuberous and thick (*Mirabilis*). They have opposite equal or unequal leave pairs. The leaves are usually simple and without stipules. They have mostly cymose inflorescence. The flowers are usually bisexual, complete, actinomorphic, and hypogynous. It has been rarely observed that the flowers are unisexual. Perianth usually consists of five united tepals and is campanulate (*Boerhavia*). Androecium usually consists of five to ten (*Bougainvillea*), three to five (*Mirabilis*), or greater number of filaments which are usually free. Anthers may be equal or not (unequal). Gynoecium consists of single carpel, the ovary is superior containing a single ovule, and they have basal placentation. An achene or nut is the most common fruit of this family. They are protected in such a way that they are enclosed in fleshy persistent perianth tube, also covered with glandular hairs. Embryo is usually curved. Perisperm is present and endosperm is absent. Pollination is carried by bees, butterflies, moths, and birds. Fleshy perianth is responsible for the dispersal of species [1].

Betalain-isolated structures are betacyanin and betaxanthin [32].

Many members of this family are a source of medicinal plants as diuretic, e.g., *Boerhavia repens*. Some members are used for ornamental purposes such as 4-o'clock (*Mirabilis*) species which are developed as ornamental orchard. Many walls are covered and fences by some *Bougainvillea* species which are usually fully fledged as hedges. One of the most common examples is *Pisonia aculeata* which can be used as hedge plant [1].

2.9 Phytolaccaceae

Phytolaccaceae belongs to a family of flowering plant. Among the different criteria of classifications, it has been recognized universally as Phytolaccaceae. Another name of this family is pokeweed family. *Rivina humilis* L. is commonly known as pigeonberry and is an important example of this family. It has been observed that it is a bushy perennial and herbaceous wild. It is distributed in colonies and has a tendency to grow on numerous shaded soil types. The maximum height of the plant has been observed 120 cm (4 ft). This species has also been distributed in tropical America and Caribbean and in Pacific and Indo-Malaysian regions naturally. High betalain pigment concentration has been observed in berries of this plant [33].

The chemical constituent which has been identified is phenolic content [34]. Betalain-isolated structures are betacyanins and betaxanthins [35].

It has been observed that members of Phytolaccaceae are good source of food, rich in nutrients, and safe to consume. One such example is *Rivina humilis* L. or pigeonberry [36]. It has been investigated that pokeberry (*Phytolacca americana* L.) which is an important example of Phytolaccaceae plays an important role as an antioxidant and antimicrobial of natural red colorant [34]. *Rivina humilis* L. has been used as natural colorant in fruits and beverage [36]. It is one of the examples of Phytolaccaceae that has been used as a source of betalain pigments which is an unutilized forest produce [36].

2.10 Portulacaceae

Portulacaceae family belongs to a family of flowering plant. Most taxonomists recognized this as Portulacaceae among different classifications. Another name of this family is "purslane family." Most commonly, these are known as sun plant, rose moss, and moss rose [37]. It has been found that family resembles with Caryophyllaceae but differs only due to calyx which consists of only two sepals. It has been widely distributed in semiarid region with highest diversity of the Southern. Some species are also found in tropical and temperate regions. Mostly they have been distributed in North and South America. Some species have been extended north into the Arctic regions. It is mostly a cosmopolitan distribution.

The family Portulacaceae has been placed in order Caryophyllales in the core eudicot clade according to the APG II system (2003). It only consists of Portulaca genus which makes the family monotypic as several genera had been removed to the Talinaceae, Didiereaceae, Montiaceae, and Anacampserotaceae in the APG III system.

There are about 440 species and 28 genera in this family and about only a single genus Portulaca. Twenty genera and about 500 species have been placed formerly. But now all the genera are removed and restricted to only one genus [1].

It consists of the following genera: *Calandrinia* consists of 120 species, *Portulaca* consists of 100 species, *Claytonia* consists of 35 species, and *Talinum* consists of 30 species. The family mostly comprises of annual or perennial herbs. But some members of this family are succulent. Cells are mucilaginous in nature which is very common and contain betalains which exhibit CAM metabolism. They have simple hair [1].

They possess herbaceous, erect, or prostrate stems. Sometimes the stems are present in the form of clusters at the end of branches. Members of this family usually possess alternate or opposite leaves. The leaves are mostly fleshy and simple. Stipules are scarious or setose but rarely absent. The family Portulacaceae has either cymose inflorescence or racemose inflorescence which is very rare.

Flowers are bisexual, actinomorphic, and solitary. Calyx consists of two sepals which are free or may be united at base and anteroposterior. Corolla consists of five petals, and infrequently it ranges from four to six. Petals are imbricate and rarely united at base. Androecium consists of many stamens; many filaments are usually free from petals or epipetalous.

Pollen may be tricolpate, polycolpate, or polyporate. Gynoecium consists of united carpels ranging from two to three. Ovary is superior and half-inferior and single chambered which contains single basal ovule. Style is usually simple but sometimes splits above, and stigma is minute. Fruit of this family is mostly a loculicidal or circumscissile capsule. Embryo is curved shape. It has been observed that perisperm is present and endosperm is absent. Pollination is favored by insects like bees, flies, and beetles. Wind and water are the sources for the dispersal of seed.

Portulacaceae family resembles with Caryophyllaceae and Basellaceae because they have common betalain component, but it has been separated from Caryophyllaceae. Morphological data assist the separation of Basellaceae and Didiereaceae.

The family Portulacaceae is monophyletic which consists of two clades. First clade consists much of *Talinum* and *Anacampseros* and its relatives *Portulaca*, *Cactaceae*, and *Talinella*. The genera which are left behind are placed in second clade [1].

The major chemical constituents which have been identified are carotene; mucilage; phenolic contents; chlorophyll; folic acid; tannins; mixture of 1-triacontanol, 1-octacosanol, and 1-hexacosanol and their respective acetates; and mixture of the steroids stigmasterol, campesterol, 3 β- D-glucosyl-sitosterol, and β-sitosterol, formerly mentioned in the root constituents [37].

Other constituents are the mixture of a pentacyclic triterpene 3-*O*-acethyl-aleuritolic acid, heneicosanoic acid, ester nonacosyl nonacosanoate, and urea 3-*O*-*β*-D-glucosyl-*β*-sitosterol; and the mixture of *β*-sitosterol and stigmasterol and long-chain hydrocarbons (dotriacontane, hentriacontane, pentatriacontane, tritriacontane) [39].

Betalain-isolated structures are two betacyanins and two betaxanthins from flower, stem, and leaf. Other betalain-isolated structures are isobetanin, betanin, betalamic acid, and immonium conjugates with tyrosine and dopamine.

Members of Portulacaceae, e.g., *P. grandiflora*, play an important role in immunomodulation because it enhances lymphocyte proliferation [37]. Many members were traditionally used for medicinal purposes such as to treat inflammatory process in general. *T. paniculatum* has been used to treat urine with bad smelling. They have also been used for the ailments of physical debility, gastrointestinal disorders, and overall weaknesses of the body. Many parts of the plants have been used for different remedies, e.g., leaves have been used in the cure of inflammatory skin diseases and superficial skin lesions such as scratches, scrapes, edemas, and minor cuts. The root portion in its powdered form has been used for the ailment of scurvy, nerve distention, and arthritis [39].

Antioxidant activity is shown by some of the members of this family such as *P. grandiflora* [37]. One of the important examples of Portulacaceae is *T. triangulare* which is very important and beneficial to plants because it gives natural colorant in foods and beverage industries [38]. Many members of this family are a source of food because they are succulent and have been used in different salads, e.g., *Montia* (Miner's lettuce) [39]. Some plants are used for ornamental purposes, e.g., rock purslane (*Calandrinia* spp.) and flameflower (*Talinum* spp.) are grown as ornamentals [1]. Portulacaceae plays an important role in medicine. One of them is used in type 2 diabetes treatments [39].

References

1. Singh, G. (2016). Plant systematics. In *An integrated approach* (3rd ed.). Boca Raton: CRC Press.
2. Impellizzeri, G., Mangiafico, S., Oriente, G., Piattelli, M., Sciuto, S., Fattorusso, E., et al. (1975). Amino acids and low-molecular-weight carbohydrates of some marine red algae. *Phytochemistry, 14*(7), 1549–1557.
3. Radfar, M., Sudarshana, M. S., & Niranjan, M. H. (2012). Betalains from stem callus cultures of Zaleya decandra LN Burm. f.-A medicinal herb. *Journal of Medicinal Plants Research, 6*(12), 2443–2447.
4. Kumar, S., Singh, P., Mishra, G., Srivastar, S., Jha, K. K., & Khosa, R. L. (2011). Phytopharmacological review of Alternanthera brasiliana (Amaranthaceae). *Asian Journal of Plant Science and Research, 1*(1), 41–47.
5. Hilou, A., Millogo-Rasolodimby, J., & Nacoulma, O. G. (2013). Betacyanins are the most relevant antioxidant molecules of Amaranthus spinosus and Boerhavia erecta. *Journal of Medicinal Plants Research, 7*(11), 645–652.
6. Cai, Y., Sun, M., & Corke, H. (2005). HPLC characterization of betalains from plants in the Amaranthaceae. *Journal of Chromatographic Science, 43*(9), 454–460.
7. Pramanik, P., Bhattacharjee, R., & Bhattacharyya, S. (2014). Evaluation of in vitro antioxidant potential of red Amaranth (Amaranthus tricolor) and green Amaranth (Amaranthus viridis) leaves extracted at different temperatures and pH. *Annals of Biological Sciences, 2*, 26–32.
8. Chong, P. H., Yusof, Y. A., Aziz, M. G., Chin, N. L., & Syed Muhammad, S. K. (2014). Evaluation of solvent extraction of Amaranth betacyanins using multivariate analysis. *International Food Research Journal, 21*(4), 1569–1573.
9. Biswas, M., Dey, S., & Sen, R. (2013). Betalains from Amaranthus tricolor L. *Journal of Pharmacognosy and Phytochemistry, 1*(5), 87–95.
10. Zhou, C., Zhang, L., Wang, H., & Chen, C. (2012). Effect of Amaranthus pigments on quality characteristics of pork sausages. *Asian-Australasian Journal of Animal Sciences, 25*(10), 1493.
11. Adhikari, R., Naveen Kumar, H. N., & Shruthi, S. D. (2012). A review on medicinal importance of Basella alba L. *International Journal of Pharmaceutical Sciences and Drug Research, 4*(2), 110–114.
12. Chatchawal, C., Nualkaew, N., Preeprame, S., Porasuphatana, S., & Priprame, A. (2010). Physical and biological properties of mucilage from Basella alba L. stem and its gel formulation. *Isan Journal of Pharmaceutical Sciences, IJPS, 6*(3), 104–112.
13. Wybraniec, S., & Mizrahi, Y. (2002). Fruit flesh betacyanin pigments in Hylocereus cacti. *Journal of Agricultural and Food Chemistry, 50*(21), 6086–6089.
14. Prajapati, P. K., Singh, S. B., & Jaiswal, S. (2014). Overview on anti-ulcer activity of Basella Alba: A therapeutic herb. *International Archive Applied Science Technology, 5*, 49–61.
15. Swati, S., & Agarwal, P. (2015). A critical review of Potaki (Basella Alba) in Ayurvedic texts with recent studies. *Ayushdhara, 2*(3), 194–198.
16. Kumar, B. R. (2016). A review on metabolic engineering approaches for enrichment and production of new secondary metabolites in Basella species. *World Journal of Pharmaceutical Sciences, 5*, 652–671.
17. Kumar, S., Prasad, A. K., Iyer, S. V., & Vaidya, S. K. (2013). Systematic pharmacognostical, phytochemical and pharmacological review on an ethno medicinal plant, Basella alba L. *Journal of Pharmacognosy and Phytotherapy, 5*(4), 53–58.
18. Wybraniec, S., & Nowak-Wydra, B. (2007). Mammillarinin: A new malonylated betacyanin from fruits of Mammillaria. *Journal of Agricultural and Food Chemistry, 55*(20), 8138–8143.
19. Szot, D., Starzak, K., Skopińska, A., & Wybraniec, S. (2015). Mass spectrometric detection of new betalains in Mammillaria flowers. *Zeszyty Naukowe Towarzystwa Doktorantów Uniwersytetu Jagiellońskiego. Nauki Ścisłe, 10*, 189–196.
20. Kobayashi, N. (2002). Contributions to betalain biochemistry: New structures, condensation reactions, and vacuolar transport. *Methods, 12*, 25.

21. Ondarza, M. A. (2016). Cactus Mucilages: Nutritional, health benefits and clinical trials. *Journal of Medical and Biological Science Research, 2*, 87–103.
22. Elmarzugi, N. A. (2016). Phytochemical properties and health benefits of Hylocereus undatus. *Nanomed Nanotechnol, 1*(1), 1–10.
23. Hamid, N. B. (2011). *Inclusion complex formation between natural dye extracted from pitaya fruit skin and B-Cyclodextrin: Kinetic and thermodynamic study*. Doctoral Dissertation, Universiti Malaysia Pahang.
24. Ortiz-Hernández, Y. D., & Carrillo-Salazar, J. A. (2012). Pitahaya (Hylocereus spp.): A short review. *Comunicata Scientiae, 3*(4), 220–237.
25. Moshfeghi, N., Mahdavi, O., Shahhosseini, F., Malekifar, S., & Khadijeh, S. (2013). Taghizadeh 5 1, 2 Master of Business Administration. *IJRRAS, 15*, 2.
26. Nadia, C., Hayette, L., Safia, M., Yasmina, M., Yasmina, H., & Abderezak, T. (2013). Physico-chemical characterisation and antioxidant activity of some Opuntia ficus-indica varieties grown in North Algeria. *African Journal of Biotechnology, 12*(3), 299–307.
27. Piga, A. (2004). Cactus pear: A fruit of nutraceutical and functional importance. *Journal of the Professional Association for Cactus Development, 6*, 9–22.
28. Jimenez-Aguilar D, M., Mújica-Paz, H., & Welti-Chanes, J. (2014). Phytochemical characterization of prickly pear (Opuntia spp.) and of its nutritional and functional properties: A review. *Current Nutrition & Food Science, 10*(1), 57–69.
29. Paiva, P. M. G., de Souza, I. F. A. C., Costa, M. C. V. V., Santos, A. D. F. S., & Coelho, L. C. B. B. (2016). Opuntia sp. Cactus: Biological Characteristics. *Cultivation and Applications, 7*(3), 1–14.
30. Abderrahim, F., Huanatico, E., Segura, R., Arribas, S., Gonzalez, M. C., & Condezo-Hoyos, L. (2015). Physical features, phenolic compounds, betalains and total antioxidant capacity of coloured quinoa seeds (Chenopodium quinoa Willd.) from Peruvian Altiplano. *Food Chemistry, 183*, 83–90.
31. Kugler, F., Graneis, S., Stintzing, F. C., & Carle, R. (2007). Studies on betaxanthin profiles of vegetables and fruits from the Chenopodiaceae and Cactaceae. *Zeitschrift für Naturforschung C, 62*(5–6), 311–318.
32. Kugler, F., Stintzing, F. C., & Carle, R. (2007). Characterisation of betalain patterns of differently coloured inflorescences from Gomphrena globosa L. and Bougainvillea sp. by HPLC–DAD–ESI–MSn. *Analytical and Bioanalytical Chemistry, 387*(2), 637–648.
33. Khan, M. I., Joseph, K. D., Ramesh, H. P., Giridhar, P., & Ravishankar, G. A. (2011). Acute, subacute and subchronic safety assessment of betalains rich Rivina humilis L. berry juice in rats. *Food and Chemical Toxicology, 49*(12), 3154–3157.
34. Mchedlishvili, N. I., Omiadze, N. T., Abutidze, M. O., Rodriguez-Lopez, J. N., Sadunishvili, T. A., Gurielidze, M. A., & Kvesitadze, G. I. (2014). Investigation of phenolic content, antioxidant and antimicrobial activities of natural food red colorant from phytolacca americana l *Fruits. Annals of Agrarian Science, 12*, 3.
35. Khan, M. I., Kumar, A., & Giridhar, P. (2016). Betalains and expression of antioxidant enzymes during development and abiotic stress in Rivina humilis L. berries. *Turkish Journal of Botany, 40*(1), 28–36.
36. Khan, M. I., Harsha, P. S., Chauhan, A. S., Vijayendra, S. V. N., Asha, M. R., & Giridhar, P. (2015). Betalains rich Rivina humilis L. berry extract as natural colorant in product (fruit spread and RTS beverage) development. *Journal of Food Science and Technology, 52*(3), 1808–1813.
37. Lim, C. K., Tiong, W. N., & Loo, J. L. (2014). Antioxidant activity and total phenolic content of different varieties of Portulaca grandiflora. *International Journal of Phytopharmacy, 4*(1), 01–05.
38. Swarna, J., Lokeswari, T. S., Smita, M., & Ravindhran, R. (2013). Characterisation and determination of in vitro antioxidant potential of betalains from Talinum triangulare (Jacq.) *Willd. Food Chemistry, 141*(4), 4382–4390.
39. Ramos, M. P. O., Silva, G. D. F., Duarte, L. P., Peres, V., Miranda, R. R. S., de Souza, G. H. B., & Vieira, F. (2010). Antinociceptive and edematogenic activity and chemical constituents of Talinum paniculatum Willd. *Journal of Chemical and Pharmaceutical Research, 2*(6), 265–274.

Chapter 3
Chemistry of Betalains

3.1 Chemical Structure of Betalain

Ecology gradually affects the biochemical composition of plant which is progressively influenced by conservationism. This observation is extensively supported when Caryophyllales is examined for its phytochemical characterization. So, anthocyanins are replaced by betalain in various vegetative parts mainly in flowers and fruits. Due to physiological features, betalain and anthocyanins are homologous, and this exploration is broadly reinforced by analysis and phytochemical identification. It was inferred that betalain has replaced anthocyanins in several parts of plants especially somatic organs chiefly in flowers and fruits.

It is very challenging to segregate betalain and anthocyanins in dissimilar classes due to correlative physiological attributes. These two compounds differ in a set of pigments comprising of 1,7-diazaheptamethin as supported by evidences. Betalains are also regarded as chromoalkaloids [1]. The range from yellow to purplish red is a chiefly observed color spectrum. The chemical diversity of secondary metabolites like betalain and related molecules is responsible for this event. Study of its chemistry began in the mid-twentieth century in Dreiding's laboratory. They were considered as anthocyanins or more specifically nitrogen-containing anthocyanins in the middle of the nineteenth century [2]. They are not only present in higher plants but unusually found in fungi as well. They are prepared in cytoplasm and then precisely stockpiled in vacuoles of fruits and flowers.

Chemically, betalain encompasses 1,2,4,7,7-pentasubstitued 1,7-diazaheptamethin in its protonated form. This pigment is known as betalamic acid having [4-(2-oxoethylidene)-1,2,3,4-tetrahydropyridine-2,6-dicarboxylic acid, thus accounting for acidic nature of betalains. The emerging research on this molecule recognizes it as an alkaloid due to ammonium analogue of betalamic acid (Fig. 3.1) [3].

E. Akbar Hussain et al., *Betalains: Biomolecular Aspects*,
https://doi.org/10.1007/978-3-319-95624-4_3

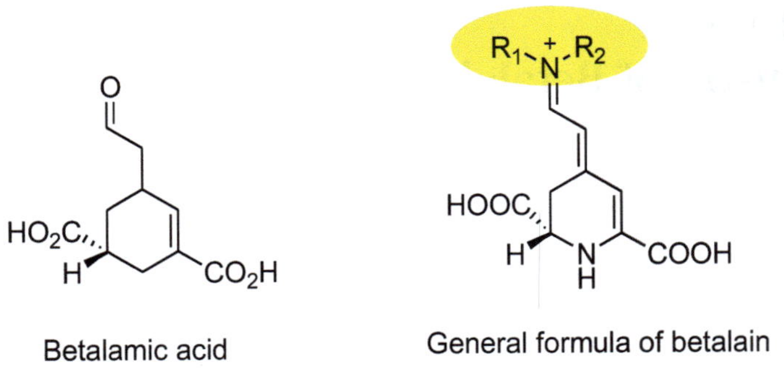

Fig. 3.1 Chemical structure of betalamic acid and betalain

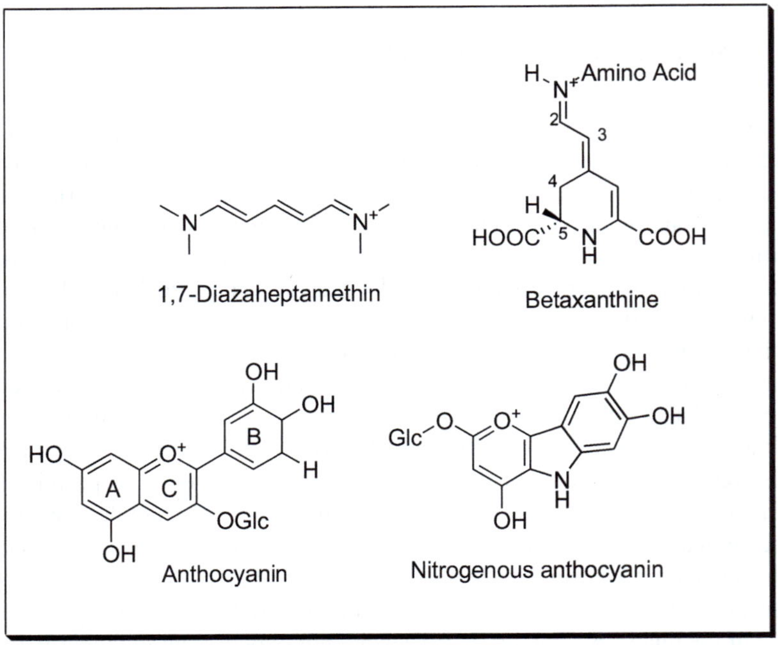

Fig. 3.2 Key building units of betalain

3.2 Key Building Units of Betalain

Structurally betalain is composed of aglycone betanidin that is linked with *β*-glycoside unit at C-5.The fundamental structure of betacyanin consists of betanidin which is formed by condensation of cyclo-DOPA (cyclo-3,4-dihydroxyphenylalanine) with betalamic acid (406 nm) [3]. In betalains, the double bonds in resonance are accountable for color array. The resonance framework due to three double bonds is liable for its lemon yellow color at wavelength 424 nm [4] (Fig. 3.2).

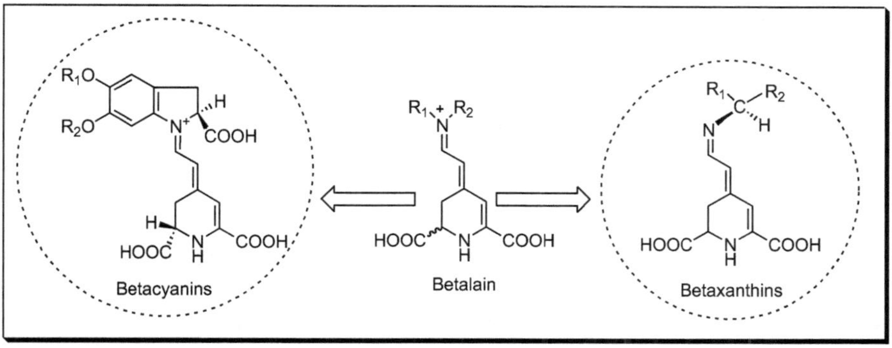

Fig. 3.3 Subclasses of betalain

The clear understanding of chemistry of this natural pigment cannot be achieved without knowing its classification into two subclasses: betacyanins and betaxanthins (Fig. 3.3).

Isolation of a distinct set of pigments such as indicaxanthin and hydrolysis of betanidin to betanin reveal the presence of 1,7-diazaheptamethin responsible for their chromas [2]. Diazaheptamethin yields derivatives (λ_{max} 340–360 nm) exhibiting 80–100 nm bathochromic shift in solution, and isolation of indicaxanthin and hydrolysis of betanidin and betanin and recrystallization prove the presence of 1,7-diazaheptamethin which is responsible for their color system in these different sets of pigments [2]. Betalains, which vary from red/violet betacyanins to yellow betaxanthins, exhibit colossal diversity of color due to acylation and glycosylation [3]. This pigment has deep violet shade because of its exceptional aromatic structure at λ_{max} 534–554 nm showing a bathochromic shift of 60–70 nm as compared to betaxanthins [4]. Betacyanin structure varies due to acyl group and additionally to sugar moieties [5]. Betalain has an interesting feature of fluorescence due to its unique atomic arrangement. As betalains are nitrogen-containing plant coloring agents, therefore they exhibit a range of colors. The conjugated double bond in the structure of betalain and its subclass members is responsible for imparting colors to plant [6].

Betaxanthins have a trait of fluorescence due to physical insinuations. But this feature is absent in betacyanins. The methylation of 1,7-diazaheptamethin via diazomethane yields derivatives (λ_{max} 340–360 nm), confirming the presence of this system in all betalain pigments which are responsible for bathochromic shifts in acidic solution. The electronic movement in 1,7-diazaheptamethin might be responsible for luminescence, but propagation of resonance toward the indole moiety causes paucity of fluorescence in case of betacyanins due to electron-deficient nature of nitrogen. Moreover, the existence of carboxyl group uplifts the fluorescence complementary with the help of indicaxanthin- and betanin-specific biosynthetic strategy [2]. Fluorescence is a significant aspect of betalain, and it is analyzed by using nanotechnology which is an emerging technique in various fields of life. The electron donating groups such as hydroxyl and aromatic groups abridge the intensity of fluorescence.

3.3 Stereoisomerism in Betalain

Another common structural feature among betalains is diastereomerism at C-2 and C-15 in betacyanins while C-11 and its corresponding carbon in betaxanthins (Fig. 3.3). The extended vulnerability of 5% citric acid in aqueous media inverts the isoform (R) into (S). On the contrary, dilute alkali action in the vacuity of oxygen converses the isomerization process [2].

It is explained that betalains are not authentic pigments, and at some point of extraction, those are artifact-generated isoforms. In natural product chemistry for structure elucidation, purification is the crucial step. This can be achieved by analytical purification techniques such as column, paper, and thin layer chromatography, which are the most utilized methods for attaining purification. Moreover, infrared and visible UV spectroscopy along with chemical and enzymatic hydrolysis, electrophoretic migration, and functional group tests describe the way of side-chain substitution [2]. Glucose unit is attached to betanidin at C-5 via *β*-glycosidic linkage to form betanin [3, 7] (Fig. 3.4).

Betacyanins seldomly occur as 2-descarboxybetanidin and often as O-glycosides of betanidin and isobetanidin, while dihydroindole unit contains hydroxyl group connected with sugar. The most prevalent sugar moiety is glucose; however other sugar moieties like xylose, galactose, rhamnose, and fructose occur less often. Ferulic, coumaric, sinapinic, and caffeic acids are hydroxycinnamic acids which give esterification with mono-, di-, or trisaccharides. Additionally, 3-hydroxy-3-methylglutaric, malonic, and citric acid are also utilized as ester analogues [3].

Isobetanidin (2S, 15R) is the C-15 epimer of betanidin (2S, 15S). They are seldom connected at 6-*O*-glucosylated, as observed in gomphrenin II, and, for the most part, at 5-*O*-glucosylated in betanin (Fig. 3.5). However, both positions are never glucosylated [3]. The plant, which belongs to the order Caryophyllales, contains betalains, while betaxanthins and betacyanins are divided as subgroup of water-soluble pigment. Betalains are conjugates of the chromophore betalamic acid that are obtained from dihydroxyphenylalanine by an oxidative 4,5-extradiol ring opening mechanism [2]. Resveratrol and anthocyanidin are known as bioactive metabolites in comparison with dopamine-derived betaxanthin (miraxanthin V) [7].

There are four groups into which betacyanins are classified (Fig. 3.6).

The widely recognized betacyanin in the kingdom of plant is betanin (betanidin-5-O-β-glucoside) (Fig. 3.7). In *Beta vulgaris* (pink beet), regular red coloration is produced from purple betanin that is a widely known betacyanin [3].

3.4 Structural Feature of Betaxanthins

Aldehyde functionality of betalamic acid when reacting with an amino acid or an amine moiety results in condensation product such as Schiff base of betaxanthins that has the same dihydropyridine structure (Fig. 3.8). *Opuntia ficus-indica* (cactus pear)

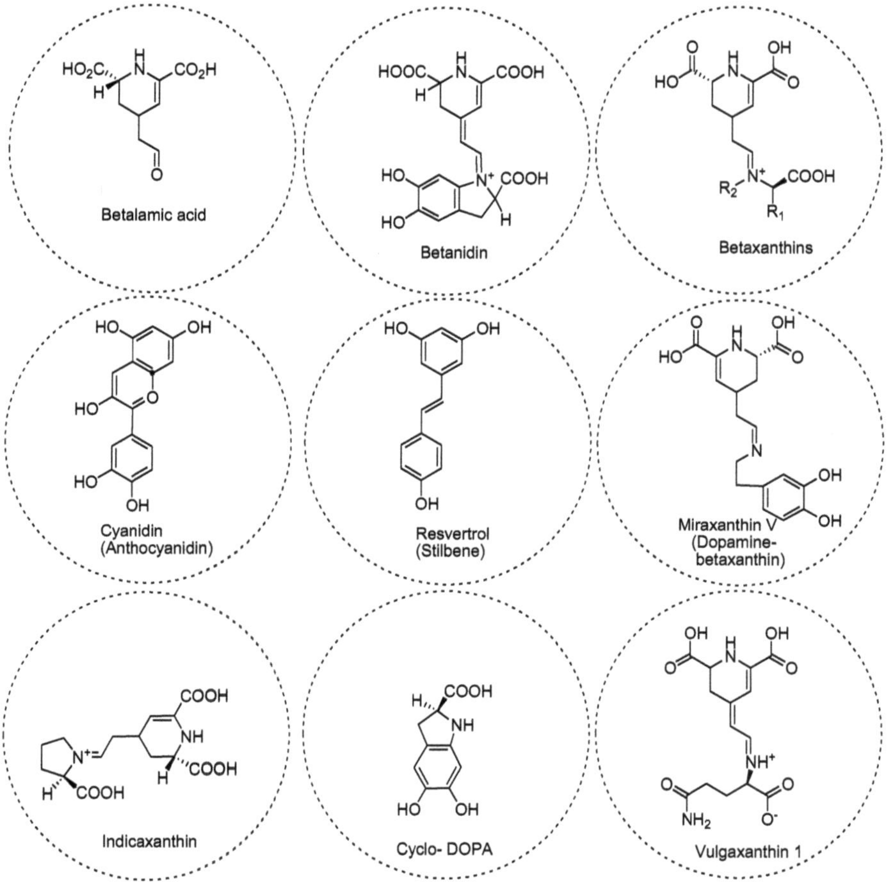

Fig. 3.4 Structures from the library of betalain

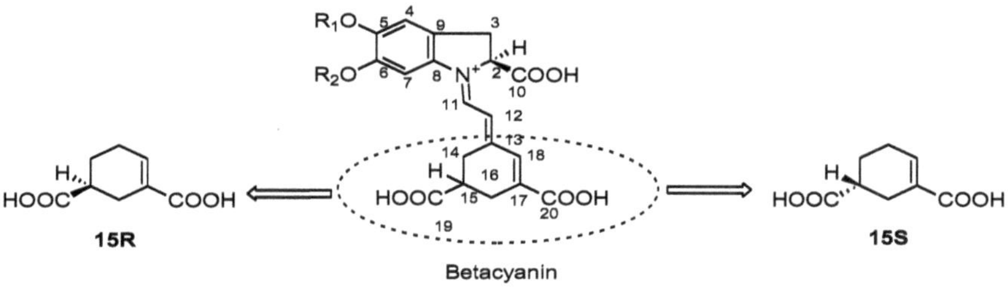

Fig. 3.5 Epimers of in betanidin

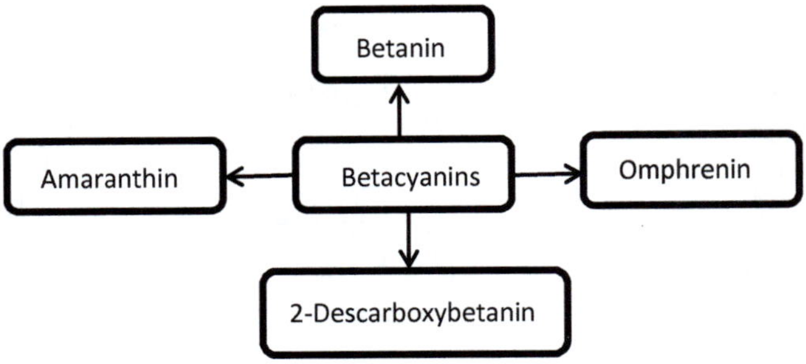

Fig. 3.6 Classification of betacyanins

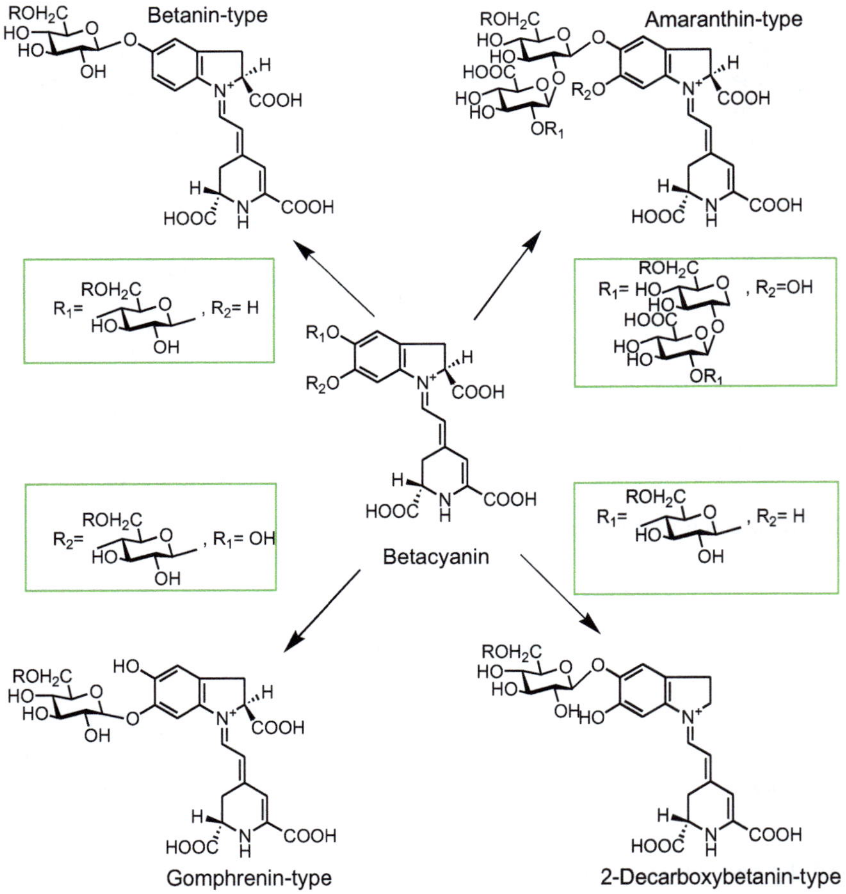

Fig. 3.7 Betacyanin derivatives

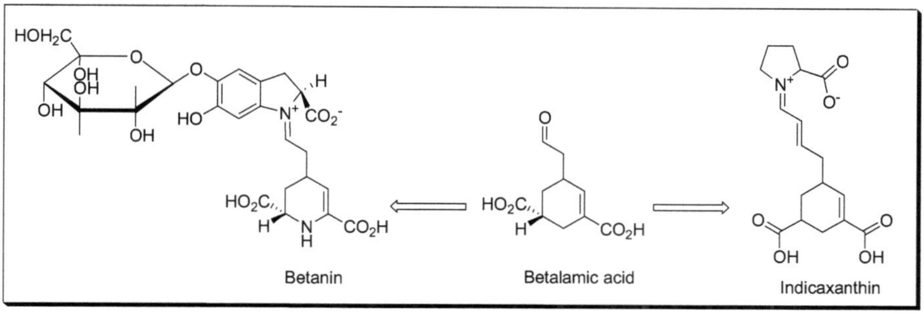

Fig. 3.8 Betalimic acid and its derivatives

and *Beta vulgaris* L. (yellow beet), which contain indicaxanthin and vulgaxanthin, respectively, are the best evidences of this observation. Dihydropyridine unit containing chiral carbons produces isomerism in betalains [3].

Two compounds have a concentration of about 4:1 ratio in beet hypocotyls. The maximum absorbance between 470 and 486 nm and the yellow-orange color or vigorous yellow color are due to this structure of betaxanthins. The yellow betaxanthins have not been reported to be glycosylated or acylated [4]. Numerous betaxanthins discovered in plants are formed by nonprotein and protein amino acids found in flora. This structural diversity of this molecule is probably due to their composition which is furnished by participation of numerous combinations of amino acids. Amino acids such as tyrosine, glycine, and tryptophanase are more frequently used as compared to other amino acids. The unique structure of various betaxanthins is a special feature depending on what family they belong to; for example, *Opuntia ficus-indica* (cactus fruit-prickly pear) carries indicaxanthin which is derived from proline [4].

3.5 Stability of Betalain

Betalains comprised of betacyanin are more stable than anthocyanins. Their color does not rely upon pH and they are stable at moderate pH. This truth gives betalains benefit than anthocyanins [8]. There are various factors that control the stability of betalain, either in increasing way or in decreasing order (Fig. 3.9).

Exogenous elements have an effect on the betalain stability as depicted in the figure (Fig. 3.10.)

Chelating agent and antioxidants such as EDTA are used for stabilization of betalains. Chelating or complex-forming metals are the elements used as stabilizing and destabilizing agents for betalain [3]. Anthocyanin regulation relies on signals which can be biotic and abiotic. Betalain pathway responds to that signals within the equal sample as anthocyanins. Living and nonliving aspects which have an effect on the betalain stability are microbes, thermal states, as well as other occasional environmental factors (Fig. 3.10).

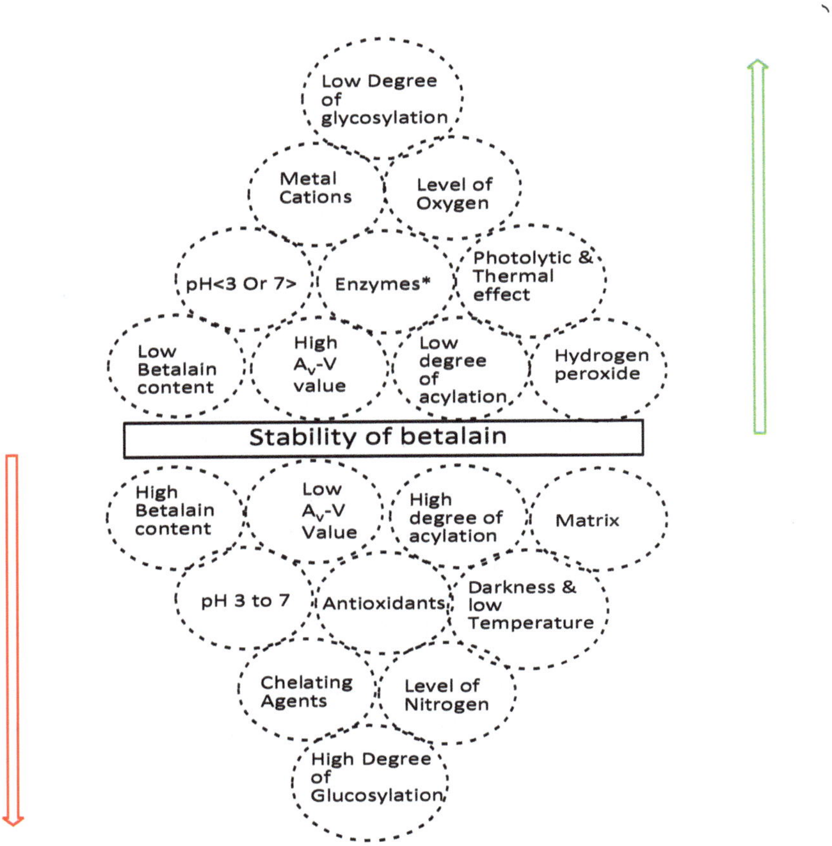

Fig. 3.9 Betalain-stabilizing and betalain-destabilizing factors

Fig. 3.10 Extrinsic factors that affect betalain stability

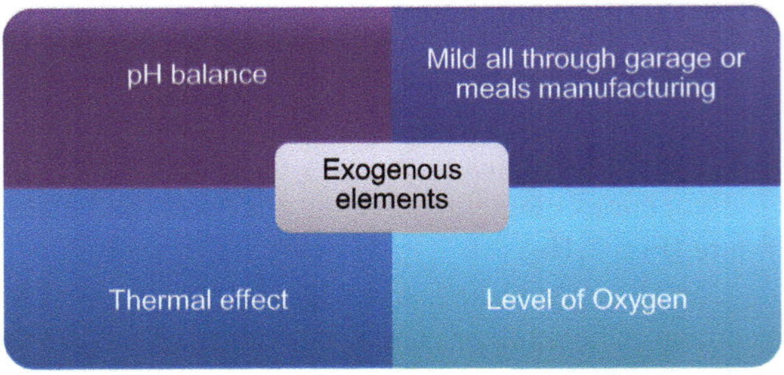

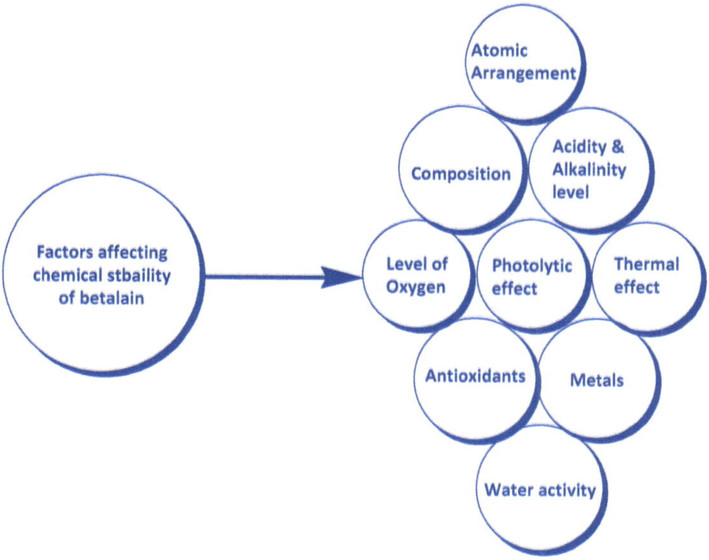

Fig. 3.11 Factors affecting chemical stability of betalain

Betalains are nutritionally beneficial due to the fact that they are scavengers of atomic oxygen species like phenyl propanoids [9]. Betalain is utilized in coloring a large collection of acids to neutral edible items as it is stabilized over a broad range of pH [7]. In comparison to betalain, anthocyanins cannot be used as color additives at pH greater than 3 due to instability. Anthocyanin is degraded using ascorbic acid, while it's far used to stabilize betalains. When experiencing water stress, *Mesembryanthemum crystallinum* transfer their mode of carbon assimilation from C-4 and C-3 to crassulacean acid metabolism (CAM), which is an advantage of betalainic plants [11].

Degradation in betalain β may additionally occur by means of special mechanisms that were precise with the aid of Herbach. Several elements, including intrinsic and extrinsic, affect the stability of betalain. It requires consideration to make sure the ideal pigment and color retention in foods containing betalains (Fig. 3.11).

Betacyanins are much more stable as compared to betaxanthins due to their structural features. The stability of this molecule maintains at temperature above 25 °C and slightly higher. Generally, betacyanins bearing glucose moiety are much more stable as compared to their aglycone counterpart. This could be rationalized on the higher potential of aglycone to be oxidized. However, substitution does not improve stability. Research has also revealed the substitution at C-6 of *O*-ester using aliphatic acids along with aromatic carboxylic acid [11] (Fig. 3.12).

Furthermore, previous research has indicated that ester formation of betacyanin with aliphatic as well as with aromatic acid at 6-*O* position enhances its stability (Fig. 3.13).

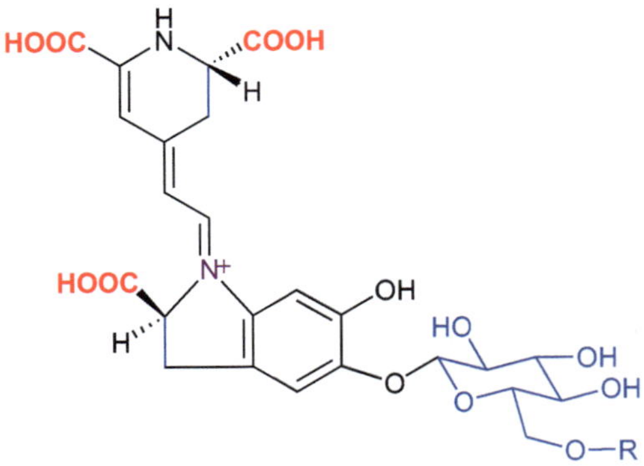

Fig. 3.12 Sites vulnerable to deglucosylation (black circled), removal of carboxyl group (red), elimination of hydrogen (blue)

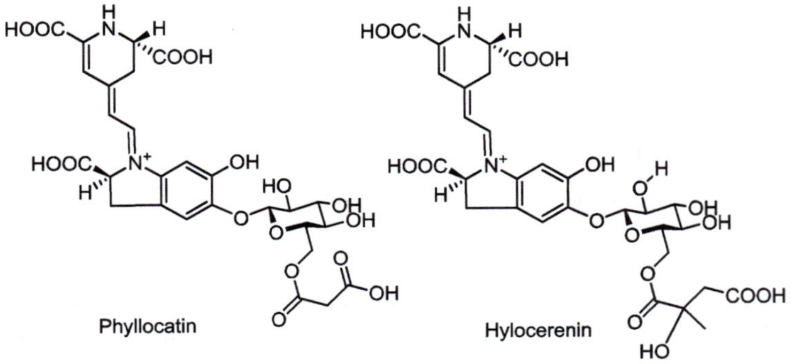

Fig. 3.13 Betalain with aliphatic and aromatic acid linkage

Thermal degradation Analytic techniques such as high-performance liquid chromatography diode array detection (HPLC-DAD) and spectrophotometry are used to monitor the thermal breakdown of betanin and betacyanins containing acyl group such as phyllocactin and hylocerenin. These methods provided remarkable consequences about thermal breakdown, such as betanin being stronger than acylated systems of betacyanins. However, the formation of products via thermal breakdown, holding excessive color, favors the color retention in phyllocactin and specifically solutions of hylocerenin. Thus, spectrophotometric technique for analysis of thermal degradation of betalain pigments is insufficient to confer the stability associated with their structure.

Endogenous enzyme action Red beetroot contains a broad range of endogenic biocatalysts such as β-glucosidases, polyphenol oxidases, and peroxidases which might be responsible for betalain degradation and discoloration if these enzymes are not properly deactivated by photolysis. The optimum pH for biocatalytic degradation of both pigments of betalain such as betacyanins and betaxanthins is pronounced to be acidic as 3.4 on pH scale. Biocatalyzed degraded products of betalain are similar to those as obtained by its thermolysis and acidic and alkaline breakdown. In betalain, betacyanins are more susceptible to degradation via peroxidases than betaxanthins, even as the latter are greater oxidized with the aid of hydrogen peroxide. The reason behind the instability of betaxanthins is the presence of catalase that is responsible for its thoroughly suppressed oxidation. The advantage of endogenic β-glucosidase activity can be taken by few attempts in order to extend the existing colors which are provided by red root beets. A bathochromic shift occurs by the removal of glucose moiety from betanin glycoside derivatives. However, the derivatives of betanin aglycones are more vulnerable and easy to transform by oxidation which consequently cause the disappearance of pink shade followed by browning [12].

Effect of alkalinity and acidity Betalains are not as vulnerable to hydrolytic breakdown as anthocyanins, although pH modification can control their changing. Betalain is incredibly strong over the wide range of pH values ranging from moderately acidic to neutral which is responsible for their utilization to less acidic food. The absorption maximum of betalain has shifted toward smaller wavelengths at pH value less than 4. On the other hand, pH value above neutral limit or alkaline directs the absorption maximum of betalain toward longer wavelengths. Moreover, for pH values excluding optimum pH limits 3.5–7, the range of visible spectra decreases. The maximum stability of betalain requires the pH range from 5 to 6. At higher pH values, hydrolytic cleavage of aldimine bond enhances, while lower pH values induce recondensation of betalamic acid with amine group of the addition residue. In acidic medium, C-15 isomerization as well as removal of hydrogen has been discovered. It is stated that breakdown of betanin in the presence of fluorescence light was three times better at pH level 3 as compared to pH level 5. However, it is revealed that mechanism of breakdown presented until now can provide a description for the shift of maximum absorption toward longer wavelength of 570–640 nm of betanin in acidic medium at pH level below 3 [13]. Despite the fact that activation strength for betacyanin degradation also depends on pH level, it declines with pH. For instance, betacyanin extract in pitaya juice acidified up to pH 4 is much less than 10% at some stage in pasteurization at 80 °C for 5 min. It is stated in previous research that high temperatures move the most useful pH for stability of betacyanin toward 6. Moreover, oxygen-free atmosphere enhances the stability of betalain in acidic medium pH (4.0–5.0) [5].

Extent of oxidation Betalains are able to react with molecular oxygen. The pigment degradation of betanin depends on the level of oxygen. Thus, betanin solutions experience less discoloration when stored in environment having low level of oxygen as compared to oxygenated environment [11]. The reason behind this is that low level of oxygen enhances the pigment to be incorporated after discoloration. It is explained that stability of betalain has been enhanced by the presence of nitrogen and antioxidant [14].

Photolytic effect The stability of betalain is also controlled by its exposure to light due to the fact that the results of light exposure are negligible under oxygen-free atmosphere [15]. It is confirmed that an inverse relation existed between betalain balance and light (2200–4400 lux). The absorption of light in UV or visible region elevates the π-electrons of the chromophore of pigment to a higher level of energy (π*) causing an increase in its reactivity by dropping the activation energy of the molecule.

Effect of metallic chelation A few metallic cations such as iron, copper, tin, and aluminum had been reported to boost up betanin degradation. Metallic-pigment complex formation can occur through bathochromic and hypochromic shifts [16]. Outcomes imply that beet juice is less vulnerable to the poor impact of metal ions as compared to solutions of pure betanin due to the presence of metal-complexing agents inside the juice. Chelating agents such as citric acid and EDTA had been mentioned to stabilize betanin toward metal-catalyzed degradation [17] (Fig. 3.14).

Antioxidants The presence of antioxidant, namely, ascorbic acid as well as isoascorbic acid, is also responsible for increasing the stability of betalain. Similarly, it is also presented that during the breakdown of ascorbic acid via hydrogen peroxide, photolytic degradation of antioxidant has occurred. Moreover, there are still disagreements regarding the substitution of ascorbic acid with isoascorbic acid [13]. It is revealed that ascorbic acid has a minor impact on stability of betalain in comparison with isoascorbic acid. However, ascorbic acid can maintain the pigment content in a better way as compared to isoascorbic acid. It is represented that phenolic antioxidants, which are responsible for discontinuation of free radical reaction, become deactivated which in turn represents that oxidation of betanin by free radical chain reaction has occurred for a short time span [14]. At some stages of food packaging and dispensation of betalain, its stability predominantly depends on temperature making it a vital factor of stability. Some researches revealed that increasing temperature enhanced the degradation rates of betalain. Breakdown of betacyanin via hemolysis in red beet and purplish pitaya as well as in betanin solution or extract becomes pronounced to comply with first-order reaction kinetics [12].

Degradation of betanin During heat processing, betanin can be degraded by isomerization and removal of carboxyl group which slowly fades the red color followed by pale brown color, sooner or later [12]. Neobetanin has been generated by removal of hydrogen from betanin which makes yellow shift possible. By using alkali and basic medium, breakdown of betanin and isobetanin has occurred which in turn produces a betalamic acid and cyclo-DOPA-5-*O*-glycoside as bright yellow and colorless, respectively. It is observed that the color of betanin depends on removal of

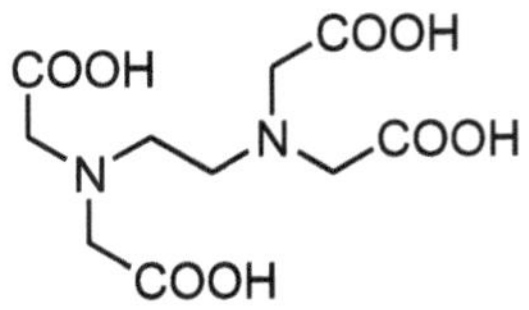

Fig. 3.14 The effective chelating agent: EDTA

carboxyl group, C-15 isomerization, and C-17 decarboxylation, which are responsible for the shift of absorption maximum toward longer wavelength from 538 to 505 nm which in turn transforms the color into orange-purple color (Fig. 3.15) [15].

Moreover, thermolysis of betanin (water /glycerol, water /ethylene glycol, and water /ethanol), at room temperature ranging from 60 to 86 °C, was performed indicating that the lowest stability of betanin has occurred. It is mentioned that the lowest stability of betanin has occurred in water/ethanol system. That thermolysis of betanin is the nucleophilic attack at aldimine bond as high electron density is being processed by oxygen of ethanol. It is also stated that breakdown of betacyanin in the presence of ethanol leads to conjugate decarboxylation. Moreover, mono-decarboxylated products of betacyanin in aqueous medium as well as in ethanol represent a unique mechanism due to different solvent media. By heating the products obtained from thermal degradation of betacyanins, unique decarboxylation ranges have been represented and identified with their analogous neo-derivatives. Following this fact, it has also explained that all ranges of mono- and bi-decarboxylated betacyanin and its derivatives are produced from heating purple beet and red pitaya preparations [11].

Influence of heating temperature on pigment loss The analysis of products, phyllocactin (malonylbetanin) and hylocerenin (three-hydroxy-3-methylglutarylbetanin), depicted that these are obtained by breakdown of betanin extracted from red pitaya juice upon heating. The most significant mechanism of breakdown in betanin was hydrolytic cleavage even as decarboxylation and dehydrogenation were predominated in hylocerenin.

Degradation of phyllocactin The products obtained by degradation of phyllocactin generated complexes regarding the removal of carboxyl group from malonic acid unit as well as betanin technology and its breakdown. By exposing the betanin to heat for a long time, the products of thermal breakdown furnished an additional double bond at C-2 and C-3.

Regeneration of betanin from degraded products As extracts and juices of betanin are stored for a short time at temperature around 10° C and at pH about 5.0, thus by

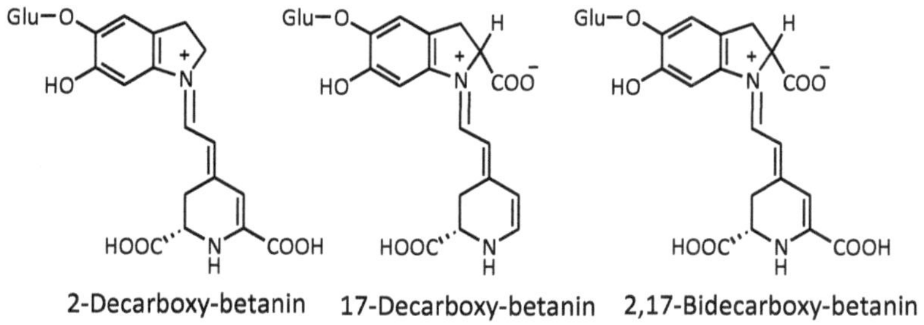

Fig. 3.15 Chemical representations of decarboxylated betanin

utilizing its foremost degraded product, betanin can be regenerated. The process of regeneration is based on hydrolytic products of betaine. The process comprises of the condensation of amine group of cyclo-DOPA-5-*O*-glycoside with the aldehyde group of betalamic acid. This regeneration speeds up when each component is reacted in solution state, aglycones can be produced from betanin by using fermentation, with betanidin / betanin ratios relying on the endogenous β-glucosidase activity of the cultivar [17].

Water activity The most important and crucial factor of stability of betalain is water activity (aw), that is, a basic unit for estimating the vulnerability of colors of betalain toward aldimine bond breaking. The impact on betalain stability can be attributed to a reduced mobility of reactants or limited oxygen solubility. It is suggested that discount advanced betanin balance, especially beneath 0.63 [18]. Moreover, it is also located that a growth of approximately one order of magnitude in betalain degradation is observed rates while accelerated from 0.32 to 0.75 [19]. In a stability analysis of encapsulated beetroot pigments, best betanin degradation came about at 0.64; this value was attributed via the authors to the reducing mobility of reactants at lower and the dilution effects at higher values [13].

Matrix effect on stability As far as matrix outcomes are concerned, storage of betanin incorporated in gelatin gels was discovered to be more favorable than in pectin gels, which ascribed to the better firmness of the previous [20]. In this light, reviews on superior stability of betaxanthins and betacyanins of their natural matrices as compared to purified solutions were viable: plant constituents like sugars, acids, and pectic materials decreased the value, thereby stabilizing betalainic pigments. Apparently, the matrix effect was discovered to be maximum by putting at pH 5, that is, the pH optimum for betanin. Furthermore, the juice matrix is changed by hydrolytic cleavage of the aldimine bond and by deviating reaction mechanisms which include decarboxylation and dehydrogenation. Therefore, utility of betalains in focused or spray-dried coloration preparations is tremendous for pigment stability; on account those matrix consequences of the fruit or vegetable juice together with water removal at some stage in processing bring about low values of the final product [15].

Betalain-stabilizing compounds Numerous food components have been defined to exert fantastic results on betalains in their herbal matrix as well as in purified pigment preparations. Supplementation with antioxidants, in particular ascorbic and isoascorbic acids, was said to embellish betalain stability by oxygen removal. But reviews on optimum ascorbic acid concentration for betalain stabilization are inconsistent, ranging from 0.003% to 0.2% and 1.0%. Conversely, decreased half-life time of betanin inside thousand ppm ascorbic acid ascribed bleaching results with the aid of hydrogen peroxide throughout ascorbic acid degradation [13]. Further disagreements exist regarding the substitute of ascorbic with isoascorbic acid. The latter was assumed to show advanced oxygen conversion due to its higher redox capability. Actually, a few studies stepped forward about betanin stability by using application of isoascorbic as opposed to ascorbic acid. However, in a latest analysis on red pitaya betacyanins, ascorbic acid was observed to yield substantially higher pigment retention than isoascorbic acid while implemented at equal concentrations. Furthermore, no pro-oxidative results of either ascorbic or isoascorbic acid will be cited in consideration that statistically tremendous pigment stabilization became received by means of elevating the antioxidant concentration from 0.1% to 1.0% [16].

Effects of citric acid and ascorbic acid Due to the fact that phenolic antioxidants and tocopherol do not show any stabilizing effect, the well-informed speculation of non-radical triggered betalain degradation. Interestingly, ascorbic and isoascorbic acid supplementation, before the heating process, has been proven to be greatly effective rather than their addition afterward. Subsequently, it may be concluded that those antioxidants do not exist longer to really improve betalain regeneration; however in addition, they prevent pigment degradation at some stage in thermal treatment [20].

Role of different compounds in betalain stabilization A variety of compounds are used for betalain stabilization such as citric acid, EDTA, metallic cationic species, etc. (Fig. 3.16). In addition to the ascorbic acid stereoisomers, chelating species have also been tested to be appropriate for betalain stabilization possibly by neutralizing the electrophilic nitrogen of betanin through attachment around the charged amino nitrogen.

Citric acid was discovered to improve betacyanin stability though being less effective than ascorbic and isoascorbic acids [21]. Complexation of metallic cations catalyzing betalain oxidation needs to, additionally, be considered. Analogously, EDTA prevents metal-catalyzed betalain degradation by pigment stabilization and complex formation with the metal ions. while another study reported that complexation of metal ions by EDTA increases betanin half-time by 1.5 times. In contrast to betanin, stepped forward vulgaxanthin I balance could not be finished by the addition of EDTA. Also preservatives and a few matrix compounds like pectin, guar

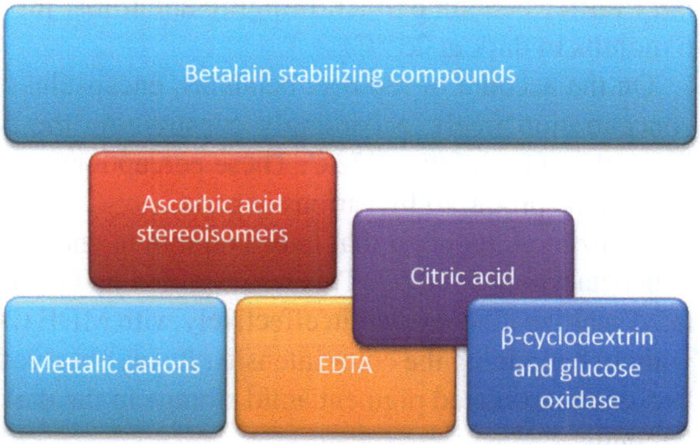

Fig. 3.16 Ascorbic acid stereoisomers

gum, and locust bean gum were proven to beautify storage stability of pink beet solutions possibly through lowering the value but to be much less influential than EDTA, ascorbic, and citric acids, respectively.

Subsequently, β-cyclodextrin and glucose oxidase had been found to be appropriate enhancers of betanin stability [17]. While the earlier may also adsorb loose water, the latter disposes dissolved oxygen, thereby contributing to pigment retention. In summary, color retention at some stages in and after the heating process of food processing of betalainic foods may be substantially expanded by exclusion of unfavorable circumstances. The disapproving conditions include metallic ions, light and oxygen, and systematic usage of unusual meal additives which include ascorbic acid (antioxidant) and citric acid (chelating agent) [20].

Encapsulation has been focused to stabilize and enhance the bioavailability and simplify the organization of polyphenols. In the sustenance industry, encapsulation utilizing spray-drying innovation is prudent and adaptable, could be worked constantly, and yields great-quality capsules. The procedure is normally utilized for creating typified phytochemicals which are dry and steady and use colorants, added substances, and flavors as nourishment. The procedure includes blending of the center material with a divider covering material. The homogenized blend of center material and divider material is sustained into a spray dryer. Through a nozzle, atomized droplets are sparged, the water in the droplets is vanished by surrounding hot air, and circular capsules of mean size reach 10–100 are acquired concerning encapsulation of betalains. It was proposed that utilization of polysaccharides, for example, pectin or guar gum as wall materials, diminished hygroscopicity in this manner expanding stability. Further investigation around there was done in the year 2000 wherein *Amaranthus* betacyanins were encapsulated utilizing maltodextrin (MDE, 10–25 dextrose proportionate) and local/altered starch as the covering operators in a food blend of 20% aggregate solids. It was observed that higher inlet temperature (>180 °C) brought about expanded drying loss (>4%) of pigment and decreased storing stability. The creators accepted a predominant storage stability of encapsulated pigments with MDE of blended dextrose proportionate (DE) taking into account the perception that there was 10–16% shading lost, which was generally less, over a four-month storage period. It was affirmed by encapsulation with MDE (10 DE) of *Opuntia lasiacantha* Pfeiffer betanin extract, bringing about lost just 14% over a capacity time of 6 months in dark at 25 °C.

On the account of red beet betalains, encapsulation ensured the colors of MDE concentration freely, accordingly constraining the degradation to just 10% amid 6 months of capacity at 27 °C. These perceptions were further upheld by comparative results produced by various experts.

Saenz et al. reported that indicaxanthin in encapsulated extract was more stable than betacyanin at 60 °C. This was certified by Gandía-Herrero et al. who encapsulated sanitized indicaxanthin effectively with MDE (20%, w/v) to settle the pigment without bargaining the color intensity. It was observed that there was no noteworthy loss of encapsulated pigment amid storage in the dark at 4 and 20 °C for more than 6 months.

Stability of the encapsulated pigment relies on the sort of shade, viz., betacyanins or betaxanthins. Likewise, these reports point to the way that encapsulation has incredible guarantee for the hygroscopic and inadequately stable betalains to augment its commercial applications through stabilization.

Pietrzkowski and Thresher (2010) protected a plan of MDE (20%) and red beet dried extract with 15% (w/w) betalains in a free-flowing item containing 5% (w/w) betalains, which was stable and very dissolvable in water.

Aside from MDE, gum arabic has been likewise utilized as an encapsulating agent. There is a plausibility of consideration of specific added substances that can give better usefulness to encapsulated betalain colors as reported by Chik and coworkers. There are wellsprings of betalains, for example, *Basella rubra* and *Opuntia ficus-indica* organic products, which contain natural sticky substances, which, if separated and/or reconstituted with colors, might go about as inherent specialists for stability and additionally esteemed options by bestowing bioactivities, for example, detoxification, assimilation, and change in hematological parameters which require further examination.

The cactus cladode adhesive blended with MDE (DE 20) could likewise be encapsulating specie with practical estimations of the adhesive.

Gelatin-MDE and capsule have additionally been tried for enhancing encapsulating proficiency, shade maintenance, storage stability, and antioxidant activity of the microparticles.

Whatever system was used to hold the most extreme color accompanied by spray drying, the prepared betalain powder ought to be adequate for consumer utility. Nevertheless, storage stability at temperatures up to 200 °C was noteworthy. Of all the drying forms, spray drying takes the succinct drying time. The drying time of droplets from the slurry is dictated by working parameters (Fig. 3.17).

In the wake of literature survey on betalain encapsulation, the ideal conditions for spray drying of MDE (DE ~ 20) + betalain/betalain-rich extract are explained in table (Table 3.1). Through this operation, the yield of encapsulated pigment can be somewhere around 90–98% and with dampness content under 5%. Dampness

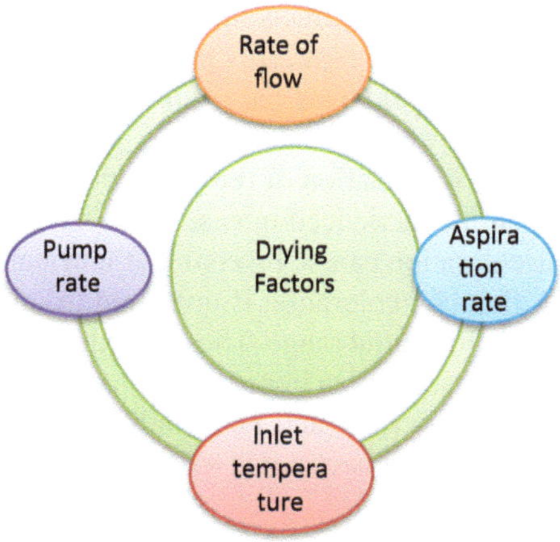

Fig. 3.17 Factors for drying of encapsulating agents

Table 3.1 Ideal conditions for spray drying of MDE and betalain

Ideal conditions for spray drying of MDE and betalain	
Food blend	~20% downright solids
Ratio of betalains and MDE	1:3
Gulf temperature	140–150 °C
Outlet temperature	65–70 °C
Feed rate	4–5 ml/min
Wind stream	5–6 l/min
Drying wind flow	40–45 m3/h or 20 psi
Value	0.6

content or quality is specifically identified with cleavage of the aldimine bond of betalains as it builds versatility of reactants, oxygen solvency, and in addition nucleophilic assault at N-1. In this way the value ought to be underneath 0.6 to enhance betalain strength. The yield and dampness substance of spray-dried betalain powder seem to diminish with expansion in inlet temperature.

By bargaining the effectiveness of the drying procedure and stability of betalain colors, Cabanes et al. had effectively encapsulated and co-encapsulated betanidin and miraxanthin V. For example, for carotenoids β-carotene and bixin, 62% and 54% yields, respectively, have separately been accounted for. Anthocyanin encapsulation yields going from 80% to 97% and 65% to 93% in the vicinity of added substances, such as acacia gum and tricalcium phosphate, have been accounted. Also, betalain retrieval in the wake of using spray drying could be enhanced by MDE blended with xanthan gum as encapsulating agent. The creators additionally analyzed that in contrast with spray drying, freeze drying was more powerful at enhancing betalains recuperation after encapsulation.

In any case, there are just a few reports related to freeze drying of betalains. This is a vital part to be investigated in light of the fact that freeze drying could be the best strategy for drying out natural product juice and foodstuffs without trading off the shade and nutritious quality given by the grid [17].

The dependability of betalains directs their scope of sustenance shading applications. Betalain extracts should be treated carefully since they are delicate to ecological conditions, especially pH, heat, light, dampness, and oxygen (Fig. 3.18). These natural components have intuitive impacts, and pigments can rapidly stain under unfriendly conditions.

The degradation of red pigment of betanin into light brown color has occurred, when it experienced intense light, air, as well as high temperature. The process of discoloration can be reversible if unfriendly conditions are just transitory.

Betalain coloration is unaltered to optimum pH range that lies in a moderately acidic to neutral range. Thus the extracts of beetroot in various types of nourishment will not stain as an immediate consequence of pH. The ideal pH for pigment of betalain, betacyanin, and betaxanthin lies in the somewhat acidic 5.0–6.0 zone. The extract of red beetroot is red colored that turns into blue color as medium becomes

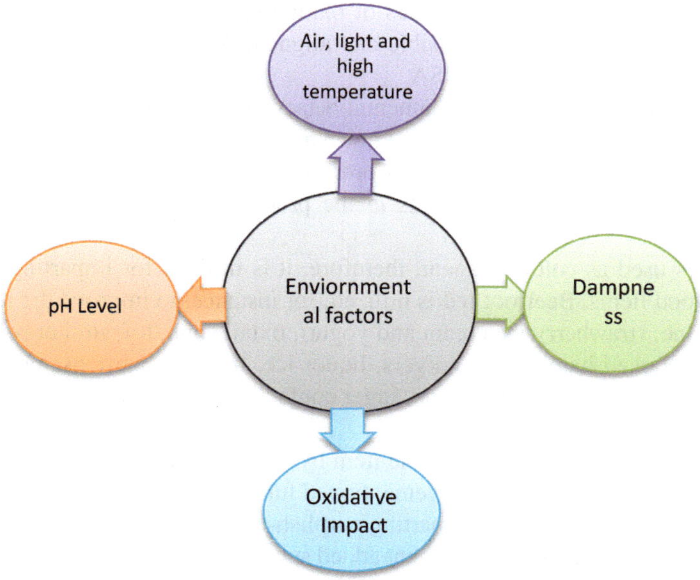

Fig. 3.18 Environmental factors

alkaline having a pH above 7. The cluster of cells in beetroot gets stained when pH level rises or medium changes toward basicity. The peeled beetroot hold its purple-red color well in acidic media, i.e., vinegar containing acetic acid. Moreover, discoloration of betalain can be achieved by heating. On slow heating, the red color changes into light brown, and discoloration will be more pronounced when mixture is hot as well as alkaline or having high pH values. The process of discoloration of betalain can be enhanced by its exposure to highly intense light.

Low dampness content in foodstuff is the most remarkable factor for stabilization of betalain. On the other hand, high level of moisture enhances the discoloration rates of betalain. Moreover, exposure to air also uplifts the rate of color fainting in eatables after some time.

Betalains respond to the oxygen all around noticeably, yet discoloration will be reversible if oxygen levels are brought down. Nevertheless, having lower stability than manufactured nourishment colors, betalain shades are broadly utilized as a part of sustenance items. Delgado-Vargas et al. listed various items containing betalain nourishment colorings. They observed that pigments of betalain are especially appropriate for their utilization in sustenance industry as they possess unique qualities. Their time span of usability is short and their production method requires low temperature, and they are easy to bundle in low moisture content and low level of light and oxygen. Moreover, they do not require humid environment for their bundling. The propensity for betalain pigments to stain under certain natural conditions has brought about them being utilized as a part of the recognition of food oxidation.

Betalain can be utilized as indicators of oxidation. The color change from red to brown on oxidation is clearly visible and aligned. Betalain strips using this technique has been licensed in the USA.

The beetroot contains the fundamental betacyanin pigment named betanin. As it is derived from beetroot, so it is also known as beetroot red. It is remarkably a very useful part of sustenance industry in order to furnish processed food items with colors instead of granting flavor as in the production of ice creams and frozen desserts.

As it is used as coloring agent, therefore, it is utilized for imparting colors to various food items. Beetroot red is utilized, for instance, to improve the redness of tomato glue, strawberry ice cream and yogurt, oxtail soup, tomato items in pizzas, sausages, cooked ham, bacon burgers, liquor ice, fruit preparations, sauces, jams, marzipan, dry powder refreshments, sugar confectionary, bread creams, and a scope of pastry items.

The extent of coloration to eatable item by using betanin is 0.1–1.5%. It is confirmed by considering all the abovementioned functions of betanin in food industry, that betanin is a basic unit for imparting purplish-red color to food items. In Europe, the system enactment for nourishment added substances is given by the Directive on Additives (1989). Extra directives spread specific added substances, including a 1994 Directive on hues utilized as a part of sustenance. Sustenance added substances, including hues, are doled out E numbers. In Europe, betanin is either recorded by name or as E162 on nourishment marks.

3.6 Coloring by Betanin Pigments

Normal nourishment colorings are experiencing a recovery inside the sustenance business. In a survey of patterns in the utilization of shading in substances, it is noticed that the utilization of actually inferred hues is growing because of enhancements in steadiness and serious concerns about synthetic food dyes from the consumers. For instance, red dye that imparts red color to foodstuffs has been denied as a result of its serious concerns.

Recent health alerts have revolved around the vicinity of the banned manufactured red colors Sudan 1 and Para Red in sustenance. Betalains have no poisonous impacts in the human body and are seen as a characteristic and safe distinct option for synthetic red colorings. Characteristic shades, for example, betalains, might turn out to be progressively utilized as a part of nourishment items. Techniques are being developed to enhance the generation of betalain in beets, through plant reproduction, cell tissue culture, and biotechnology. Notwithstanding expanding the amount and nature of betalains, a definitive point is to enhance the strength of betalain particles in nourishment items [23].

3.7 Development of Stabilization of Betalain during Dispensation and Packing

The color or shade of extract and juice of purple pitaya has been analyzed at the stage of packing and collection in order to confirm its appropriateness for common utilization.

It has been proved from previous research on red beet that hue with holding power of betalain has elevated after its dispensation and its content. Considering that early studies on red beet had proven that betalain regeneration after processing elevated overall color retention, the betalain content improvement of the acquired juices becomes registered over 72 h at 4 °C. Completion of betacyanin regeneration was achieved after 24 h and considered essential to maximize pigment yield [24]. Redeveloped samples that underwent through heated atmosphere possess almost 10% color alteration, while unheated extracts showed 3% remarkable increase in pigment intensity. It is revealed that previous research focusing on the factors stabilizing the ordinary food coloring agents has been contradictory; organic acids including citric, ascorbic, and isoascorbic have been brought to juices and pigment arrangements from 0.1 to 1% previous to heating. After high-temperature short-time (HTST) treatment at semicommercial scale, up to 2–3% of the initial betacyanin content was retained. Consequently, viability of pitaya juice was to be taken into consideration if good enough stabilization measures had been carried out to strongly enhance average pigment yield.

Redevelopment and stabilization of pigment have fluctuated between pH 4 and 6. However the analysis has focused on pH less than 6 as it is applicable for industrial methodologies.

It is important to consider that purified pigment samples without matrix had been ineffectively stabilized than samples of non-purified extracts, and a dosage of 1% ascorbic acid was found to noticeably be less effective in betacyanin degradation.

Almost 70% of primary betacyanins are not affected when kept under intense light or in the shade, while such conditions have an effect on ascorbic acid [25]. In assessment, pigment losses about 60 and 90% upon darkish and light storage at 20 °C, respectively, have been registered without ascorbic acid addition. Qualitative color change has been monitored without any difficulty as compared to quantitative findings by utilizing DE* technique. This technique included all chromatic elements such as lightness L*, inexperienced-redness a*, and blue-yellowness b*.Yellow-orange cactus pear, organic acids such as tartaric acid, oxalic acid, and malonic acid bogged down betalain degradation upon thermal degradation. Betaxanthins have been notably showing greater stability at pH 6 instead of pH 4, whereas the pH balance of betacyanins relied on the respective acid carried out [23]. The most expected effects have been received with 0.1% isoascorbic acid at pH 4 and 0.1% citric acid at pH 6. At pH 4, the half-life period for indicaxanthin had been expanded via 0.1% isoascorbic acid dosage from 78.8 to 126.6 min, 31.4 to 46.5 min, and 13.4 to 21.7 min at 75 °C, 85 °C, and 95 °C, respectively [23].

Furthermore, stabilization of betalain preparations without matrix ingredients becomes less powerful in comparison to its extract. Subsequently, a matrix index turned to express the ability of diverse organic acids to improve pigment stability. Ultimately, regeneration of betaxanthins, though at first it was not taken into consideration, was discovered to be a vital thing in maximizing pigment yield [25]. Betaxanthin regeneration without additive gets higher at pH 6 up to 6% as compared to most effective 2.5% at pH 4, while typical shade retention became pleasant by means of adding 0.1% isoascorbic acid at pH 4 or 0.1% citric acid at pH 6 before heating. Upon cactus pear juice processing at pilot-plant scale, vulgaxanthin I being the fundamental compound in red beet and Swiss chard changed into less stable than indicaxanthin [21].

This lent guide to the reality that cactus pear may be regarded as betalain pigment plant. During storage for up to half a year, pigments were once more first-rate covered, if juices were stabilized with 0.1% isoascorbic acid. The most exceptional change became registered in the course of the primary month, being much less said afterward. The half-life period of betaxanthins and betacyanins might be extended up to 2.6 and 3.6 months by the addition of 0.1% isoascorbic acid, while without addition of any additive, the half-life had been around 1 month only [22]. The impact of stabilization became less reported for the duration of storage under light storage when half-lives of 0.8 and 1.3 months had been performed for betaxanthins and betacyanins, respectively. This research on purple pitaya and orange-yellow cactus pear validated that the choice of the best additive on the proper attention will depend both on pigment kind (betacyanins and betaxanthins) and the components of the respective betalain supply. Subsequently, the stabilization strategy requires to be adjusted for each product [25].

Metal ions of Hg^{+2} and Cu^{+2} form complex with betalains, and in reaction to Cu^{+2}, Fe^{+2}, Co^{+2}, and Mo^{+2}, betalain biosynthesis is upregulated. In metallic stress resistance, useful position of betalains has never been evaluated. In addition, it is not known whether betalains identified in leaves are included in repelling herbivores or not. Regarding dry season, saltness stress reaction, and oxidative stress, betacyanins are great match for anthocyanins in leaves and other vegetative organs [11].

3.8 Bioavailability

Bioavailability offers us an idea about the useful human fitness properties of a molecule, which defines the active substance fraction while absorbed orally. *B. vulgaris* root and *O. ficus-indica* fruit are used to decide bioavailability of betalains by the usage of their extracts. Bioavailability is affected by meal processing and its source, the matrix of food, and the structure of betalains [10].

Due to the presence of conjugated double bonds in 1,7-diazaheptamethin structure and the phenolic characteristic of betalamic acid, all betalains revealed absorption maxima in both visible and UV regions, respectively. A few reviews have explained the fluorescence of betalains as well as fluorescence related to betaxanthins but not with betacyanins. As stated earlier, betacyanins are not capable of characteristic luminescence due to their structural configuration. The constant electronic movement in 1,7-diazaheptamethin is probably responsible for fluorescence, but on extension to indole moiety as in the case of betacyanins, the fluorescence is misplaced.

Electron-donating groups reduced the intensity such as aromatic ring and hydroxyl group; on the other hand, carboxyl group which is an electron-withdrawing group elevated the fluorescence intensity [7].

References

1. Zryd, J.-P., & Christinet, L. (2003). *Betalain pigments*. Université de Lausanne, CH 1015 Lausanne, Switzerland, (pp. 1–25).
2. Khan, M.-I., & Giridhar, P. (2015). Plant betalains: Chemistry and biochemistry. *Photochemistry, 117*, 267–295.
3. Tuba, E., Anika, W., Valerie, B., Schini, K., & Gerald, R. (2015). Betanin – A food colorant with biological activity. *Molecular Nutrition & Food Research, 59*, 36–47.
4. Davies, K. (2009). *Plant pigments and their manipulation, Annual Plant Reviews* (Vol. 14). Boca Raton, FL: CRC Press LLC.
5. Delgado, F. V., Jimenez, A. R., & Lopez, O. P. (2000). Natural pigments: Carotenoids, anthocyanins, and betalains — Characteristics, biosynthesis, processing, and stability. *Critical Reviews in Food Science and Nutrition, 40*(3), 173–289.
6. Castellar, R., Obón, J.-M., Alacid, M., & Fernández-López, J. A. (2003). Color properties and stability of betacyanins from Opuntia fruits. *Journal of Agricultural and Food Chemistry, 51*, 2772–2776.
7. Fernando, G.-H., Josefa, E., & Francisco, G.-C. (2016). Biological activities of plant pigments betalains. *Critical Reviews in Food Science and Nutrition, 56*, 937–945.
8. Yoshikazu, T., Nobuhiro, S., & Akemi, O. (2008). Biosynthesis of plant pigments: Anthocyanins, betalains and carotenoids. *The Plant Journal, 54*, 733–749.
9. Gregory, J.-H., & Alan, L. (2015). Pigments in fruits and vegetables. Genomics and dietetics. Chapter 6. In *The betalain secondary metabolic network* (pp. 127–140). Springer.
10. Pavokovi, D., & Krsnik, R. M. (2011). Biotechnological production of betalains. *Food Technology and Biotechnology, 49*(2), 145–155.
11. Kirsten, M.-H., Florian, C.-S., & Reinhold, C. (2006). Betalain stability and degradation—Structural and chromatic aspects. *Concise Reviews/Hypotheses in Food Science, 71*(4), 41–50.
12. Henriette, M., & Azeredo, C. (2009). Betalains: Properties, sources, applications, and stability – A review. *Food Science and Technology, 44*(12), 2365–2376.
13. El Hassana, G. (2011). Betalain: A particular class of antioxidant pigment. *Natural Product Communications, 6*(10), 1425–1431.
14. Luxsika, N., Sakamon, D., & Naphaporn, C. (2015). Natural colorants: Pigment stability and extraction yield enhancement via utilization of appropriate pretreatment and extraction methods. *Critical Reviews in Food Science and Nutritions, 57*(15), 3243–3259.
15. Herbach, K.-M., Stintzing, F. C., & Carle, R. (2004). Impact of thermal treatment on colour and pigment pattern of red beet (*Beta vulgaris L.*) preparations. *Journal of Food Science, 69*, 491–498.
16. Attoe, E.-L., & von, E.-J.-H. (1985). Oxygen involvement in betanine degradation: Effect of antioxidants. *Journal of Food Science, 50*, 106–110.
17. Florian, C.-S., & Reinhold, C. (2007). Betalains emerging prospects for food scientists. *Trends in Food Science & Technology, 18*, 514–525.

18. Kearseley, M.-W., & Katsaboxakis, K. Z. (1980). Stability and use of natural colorant in food. *Journal of Food Technology, 15*, 201–514.
19. Cohen, E. S., & Saguy, I. (1983). Effects of water activity and moisture content on stability of beet powder pigments. *Journal of Food Science, 48*, 703–707.
20. Stintzing, F.-C., Kugler, F. C., & Conrad, R. (2006). First 13C-NMR assignments of betaxanthins. *Helvetica Chimica Acta, 89*, 1008–1016.
21. Mohammer, M.-R., Stintzing, F. C., & Carle, R. (2006). Cactus pear fruits (Opuntia spp.) a review on processing technologies and current uses. *Journal of the Professional Association for Cactus Development, 8*, 1–25.
22. Wybraniec, S. (2005). Formation of decarboxylated betacyanins in heated purified betacyanin fractions from red beetroot (*Beta vulgaris L.*) monitored by LC-MS/MS. *Journal of Agricultural and Food Chemistry, 53*, 3483–3487.
23. Pasch, J.-H., von, E., & H, J. (1979). Betanine stability in buffered solutions containing organic acids, metal cations, antioxidants, or sequestrants. *Journal of Food Science, 44*, 72–81.
24. Mikova, K., & Kyzlink, V. (1985). Effect of glucose oxidase on the color stability of red beet concentrate. *Journal of Chemical Technology, 58*, 9–15.
25. Cai, Y., & Corke, H. (2001). Effect of postharvest treatments on Amaranthus betacyanin degradation evaluated by visible/near-infrared spectroscopy. *Journal of Food Science, 66*, 1112–1118.

Chapter 4
Biosynthesis of Betalains

4.1 Introduction

Betalains, the secondary metabolites, are originated from shikimic acid as well as from tyrosine amino acid by the action of arogenate dehydrogenase enzyme [1]. Basic structure of betalains is formed by bonding of phenyl group to an *n*-propyl side chain forming a three- to six-carbon building block.

Biosynthesis from tyrosine has incomplete exploration, and only the activity of few enzymes in biosynthetic pathway is identified [2]. Due to the involvement of several steps in the fabrication of betalamic acid and obtainment of insufficient yields of final desired products, chemical synthesis of betalains is not practicable for pigment production commercially [3]. For determining the structure and biosynthetic pathway of betalains, betanin (the main pigment of red beet) has been used as a model [4]. Two plant families, the Caryophyllaceae and Molluginaceae of the order Caryophyllales, contain members that produce anthocyanins instead of betalains. Apart from plants, certain genera of *Basidiomycetes* also synthesize betalains.

Different species show different types of betalain production; they may be located only in unripe organs such as anthocyanins or only in mature organs or preserve for organ's whole lifetime; they may be synthesized only in reproductive organs such as fruits and flowers of several cacti, or they may be present in both reproductive and in vegetative structures such as the flowers and leaves of ice plants. Betalain biosynthesis takes place in the cytoplasm, and then these are stored in vacuoles located primarily in fruits and flowers and in plant vegetative tissues occasionally [3].

Its biosynthesis in subcellular organs of plants is not yet clear. Few studies on tyrosine feeding reveal that betalains are synthesized in cytoplasm and then transferred to vacuoles just like to anthocyanin mechanism of intracellular translocation. The detection of cytoplasmic DOD and cytochrome P450 which are involved in betalain biosynthesis shows the worthiness of this hypothesis. However, there is something to be discovered about it by elucidation of intracellular mechanism of biomolecular transport involving tyrosine and DOPA to cytoplasm from chloroplasts and import of betalain vacuole [5].

E. Akbar Hussain et al., *Betalains: Biomolecular Aspects*,
https://doi.org/10.1007/978-3-319-95624-4_4

An evolutionary junction grounded on a distinctive dioxygenase enzyme responsible for aromatic ring cleavage of the biosynthetic pathway is evident by the correspondent pigments present in plants and fungi. The betalain biosynthetic route needs to be entirely discovered unlike other pigments of plants [6].

The foundation of currently accepted betalains was proposed as a biosynthetic pathway by showing that unlike other amino acids, incorporation of radiolabeled DOPA into betanin produced indicaxanthin 90% of the radioactivity on reaction with proline. It was concluded that DOPA is an effective forerunner of betalamic acid. After feeding radiolabeled DOPA, similar results were obtained on indicaxanthin biogenesis.

An inclusive apprise of the different betalain biogenesis reports was collected in 2003. More recently, in the last decade, advances in biogenesis of betalains were illuminated by Gandia-Herrero and Garcia-Carmona [5]. Another detailed biosynthetic pathway of betalains was proposed by Han and coworkers as well as Strack and his coworkers [3]. However, the scheme of betalain biogenesis is still a black box. Although in the current era, the main ladders of route have been defined, there is a possibility of multiple side reactions [6]. Recently, from *Amanita muscaria*, a notorious mushroom toadstool, the characterization of betalain biogenesis is made. *Amanita muscaria* belongs to basidiomycetes that piles up betalains in its cap, and the regulation of its biosynthetic development is considered much [7].

4.2 Biosynthetic Pathway

Biosynthesis of betalains is categorized into two types:

I. Biosynthesis from tyrosine (classic pathway)
II. Biosynthesis from tyramine (newly described pathway) [8]

4.2.1 Biosynthesis from Tyrosine

In this betalain biosynthetic pathway, tyrosine (Tyr), an amino acid, serves as the starting material [9].

4.2.1.1 Tyrosine Accumulation

Biosynthesis of betalain starts with the massive accumulation of tyrosine [7]. In plants, L-tyrosine is derived from arogenic acid that remained unknown in betalain papers [10]. In shikimic pathway, betalains originate from arogenate as secondary metabolites via a dehydrogenase enzyme; arogenate is transformed into an amino acid tyrosine [7] which is the forerunner of L-DOPA biosynthesis [10].

4.2.1.2 Hydroxylation of Tyrosine to L-5,6-Dihydroxyphenylalanine

Hydroxylation of tyrosine occurs by tyrosine hydroxylase activity (TOH) resulting in DOPA production [5]. From two molecules of tyrosine, synthesis of fundamental dihydropyridine core takes place in betalains, by which bimolecular formation of L-5,6-dihydroxyphenylalanine (L-DOPA) occurs [7]. In large number of plant families, DOPA is an important metabolic product which accumulates largely in few species of Leguminosae, e.g., *Vicia faba* [4].

4.2.1.3 Fates of DOPA

Conversion of DOPA occurs by three routes later on:

I. Its cleavage and opening reactions into 4,5-seco-DOPA may take place which are catalyzed by DOPA 4,5-dioxygenase (DOD).
II. Enzyme tyrosinase (EC1.10.3.1) can oxidize it to dopaquinone.
III. It may undergo decarboxylation in a DOPA-decarboxylase-catalyzed reaction leading to the formation of dopamine [3].

These three reaction sequences are described below.

4.2.1.4 Formation of Betalamic Acid

Ring opening of L-DOPA takes place between carbons 4 and 5 [10] by extradiol aromatic ring cleavage activity [5, 9] of 4,5-DOPA-extradiol dioxygenase; thereby 4,5-*seco*-DOPA is produced which serves as an intermediate and is also present in nature 10].

4,5-*Seco*-DOPA rearranges itself to betalamic acid without any enzyme [5]. Spontaneous intramolecular condensation of the amine and the aldehyde group of 4,5-*seco*-DOPA occurs in order to produce betalamic acid [10] which serves as the basic core structure of betalains [5].

In fungi, betalamic acid is produced by a fungal specific enzyme accompanied by muscaflavin, a related pigment found solely in fungi [6]. Along with plant 4,5-*seco*-DOPA product, an intermediate 2,3-*seco*-DOPA is produced by activity of an additional enzyme 2,3-DOPA-extradiol dioxygenase [11]. In plants, the enzymes extracted from *B. vulgaris*, *M. jalapa*, and *P. grandiflora* have been functionally recognized to have the activity of 4,5-DOPA-extradiol-dioxygenase enzyme [6].

Betalamic acid is actually a yellow, fluorescent compound that is essential to produce red and yellow betalains [12]. The betalamic acid produced may undergo condensation reaction with the imino group located in cyclo-DOPA to synthesize the red-purple betacyanin pigments, or it may condense with amino or imino moiety of amino acids to produce the yellow betaxanthin pigments [2].

4.2.1.5 Condensation of Betalamic Acid

Betalamic acid undergoes condensation with different moieties in order to produce betaxanthins, portulacaxanthin II, and dopaxanthins.

4.2.1.6 Betaxanthin Formation

Aldimine bond formation takes place [3] when aldehyde group of betalamic acid is condensed with an amino acid amine group spontaneously to produce the related imine [6], thereby producing betaxanthins such as indicaxanthin or vulgaxanthin [3, 9]. To produce semisynthetic betalains, this reaction can be introduced in vitro [6]. Schliemann and his coworkers reported that stereoselectivity as well as specificity cannot be observed when betalamic acid forms aldimine bond with amino acids [5].

4.2.1.7 Portulacaxanthin II Formation

If back condensation occurs between betalamic acid and tyrosine, tyrosine-betaxanthin (portulacaxanthin II) is produced which is the starting molecule of the biosynthetic route. This compound is present in *Portulaca* sp. and other species [3].

4.2.1.8 Dopaxanthin Formation

Portulacaxanthin II, by the catalytic activity of tyrosinase, is converted to dopaxanthin. L-DOPA condensation with betalamic acid also produces dopaxanthin which is converted by tyrosinase to dopaxanthin-quinone.

4.2.1.9 Formation of Cyclo-DOPA (5,6-Dihydroxyindole-2-Carboxylate)

When reducing agent is not present, the transformation of L-DOPA to *o*-DOPA-quinone takes place by tyrosinase enzyme using molecular oxygen. The amine group of *o*-quinone attacks intramolecularly as a nucleophile on the ring later on. As a result of which, spontaneous cyclization occurs that produces a molecule of leuko-DOPA-chrome also called cyclo-DOPA [6]. L-DOPA may also be converted to cyclo-DOPA by the activity of cytochrome P450 [13] as shown in *B. vulgaris*. The mechanism of cyclo-DOPA production is not identified completely. There may be two possibilities: oxidation of one-electron (stepwise electron transfer) as well as classical oxidation. In both cases, the intermediate oxidation product is DOPA-quinone, before the intramolecular nucleophilic attack occurs on the ring. However, due to the deficiency of evidences for the mechanism of DOPA oxidation for the cytochrome P450 subfamily, this conversion is given distinctly in biosynthetic scheme.

For the synthesis of betalamic acid, L-DOPA should be accumulated, which requires end of oxidation reaction of L-DOPA which is catalyzed by the tyrosinase due to its diphenolase activity. This reaction involves the whole consumption of synthesized L-DOPA as a substrate and is important for forming a part of the betacyanin structure. The back transformation of *o*-DOPA-quinone to L-DOPA requires ascorbic acid or an equivalent reducing agent. This ensures the availability of L-DOPA for further steps. Betalain-accumulating vegetable material stores a significant amount of ascorbic acid.

4.2.1.10 Oxidation of Cyclo-DOPA

Leuko-DOPA-chrome is unstable as indicated by the proposed reactions; it is converted spontaneously to DOPA-chrome by oxidation with the consequent back reduction of DOPA-quinone molecule to L-DOPA. This shows that, when reducing agent is not present, there is an excess of DOPA-chrome instead of cyclo-DOPA. In addition, further development of DOPA-chrome takes place for the formation of brown polymers which are responsible for enzymatic browning. Leuko-DOPA-chrome can be obtained from L-DOPA or L-tyrosine by getting the DOPA-chrome phase when reducing agents are not present and brown polymers are not formed and then allowing it to react with such a reducing agent which is capable to transmute it to leuko-DOPA-chrome [6].

4.2.1.11 DOPA Decarboxylation

Decarboxylation of DOPA occurs by means of an enzyme DOPA decarboxylase, thereby producing dopamine [10]. Conversion of DOPA into dopamine via decarboxylation is needed for dopamine-derived betalains such as miraxanthin V or the 2-descarboxy-betanin. The presence of decarboxylase enzyme responsible for the transformation of DOPA into dopamine is ascertained by protein extract obtained from red beet by which synthesis of dopamine-derived betacyanins is supported [4]. Decarboxylated betalains are present in plants, or they may be evident by in vitro cultures.

4.2.1.12 Formation of 2-Descarboxybetanidin

By the activity of tyrosinase and then by cyclization, the 2-descarboxy-cyclo-DOPA is produced from dopamine which undergoes condensation with betalamic acid through imine bond and yields 2-descarboxy-betanidin. Condensation of betalamic acid with the decarboxylated biogenous amines dopamine and tyramine may able to synthesize dopamine-betaxanthin and tyramine-betaxanthin, from which 2-descarboxy-betanidin is formed by the activity of tyrosinase [10].

4.2.1.13 Formation of Betanidin

Betanidin is a significant precursor for the betacyanin production [13] is formed via condensation of cyclo-DOPA [14] and betalamic acid [15] in the same manner in which imine bond formation takes place [10].

However, it is not clear yet if the betalamic acids participating in the production of betaxanthins and betacyanins arise from the same pool or not. It seems that betanidin may come from betaxanthins in the absence of cyclo-DOPA (when betalamic acid is back condensed to tyrosine, portulacaxanthin II is formed, which is converted by the activity of tyrosinase to dopaxanthin and then to dopaxanthin-quinone followed by the formation of betanidin). Up till now, betanidin biogenesis through deglucosylation has been suggested only in red beetroots. But due to the mutual existence of tyrosine, DOPA, and oxidizing enzymes, betanidin biogenesis in other plants is equally associated by betalamic acid condensation with cyclo-DOPA and reaction sequence from back condensation of betalamic acid to betanidin described above. Explanation of accessibility of betanidin content for co-accumulation of its several products, for example, betanidin-6-O-glucosides as well as betanidin-5-O-glucosides in *G. globosa* petals and ulluco tubers, is obtained by this supposed material [5].

By taking into account the role of dopaxanthin and tyrosine-betaxanthin, biogenesis of betanidin can also be explained. Betaxanthin is synthesized when L-tyrosine is condensed with betalamic acid. Tyrosinase catalyzes the conversion of betaxanthin pigment by its monophenolase activity to the second pigment dopaxanthin. Dopaxanthin may also serve as substrate for tyrosinase and is converted to dopaxanthin-quinone. A reducing agent is required for the maintenance of dopaxanthin in the presence of tyrosinase which reverts it to *o*-quinone which is an initial pigment just like the case of free L-DOPA. In plants, high levels of ascorbic acid are present in dopaxanthin-accumulating flowers which protect them from oxidation [16]. In case when reducing agent is not present, an intramolecular nucleophilic attack is promoted by dopaxanthin-quinone, which results in its cyclization to betanidin analogous molecules, with same properties and molecular weights, as the case with the reactions that occurred to L-DOPA. Unlike the exclusive leuko-DOPA-chrome which is obtained from free L-DOPA, the additional structure of dopaxanthin extends the nucleophilicity due to resonance, thus producing several cyclic molecules. It is not yet determined under which conditions betanidin is produced from dopaxanthin-quinone, but the route is valid apparently, and the biogenesis of betanidin is favored in vivo. Alternative approach to this division is the condensation between L-DOPA and betalamic acid to generate dopaxanthin.

4.2.1.14 Transformation of Betanidin to Betanidin-Quinone

Betanidin is converted to betanidin-quinone by the radical mechanism action of peroxidase. Decarboxylated and dehydrogenated derivatives of betanidin-quinone may be synthesized which depends on the pH [6].

4.2.1.15 Acylation, Glucosylation, and Other Modifications

Mostly, the secondary compounds of plants are produced first as basic structure, and then these are modified by glucosyltransferase enzymes with sugar moieties and by acyl transferase enzymes with aromatic or aliphatic acyl moieties. For betalain biosynthetic pathway, two routes are described based on experiments and early phytochemical experiments. In the first route, glucosylation of betanidin takes place, which resembles biosynthetic pathways of other secondary products of plants. In the second route, glucosylation of cyclo-DOPA occurs by which betalamic acid condenses with cyclo-DOPA molecules modified with sugar and acyl groups. It is still a question which pathway is favored for the in vivo synthesis of betacyanins.

Ground for the first route was established by experimentation in which fruits of tumid prickly pear were incorporated by exogenously supplied betanidin which was radiolabeled. This route was also supported by biochemical data having the determination of glucosyltransferase action toward betanidin in unrefined enzyme generated from living stone daisy-cultured cells *Dorotheanthus bellidiformis*. From living stone daisy cell cultures, [UDP-glucose: betanidin 5-O-glucosyltransferase (B5GT) and UDP-glucose: betanidin 6-O-glucosyltransferase (B6GT)] which are two region-specific enzymes and are responsible for glucosylation of betanidin were purified, and their detailed enzymatic characteristics were described. These enzymes can take flavonoids and betanidin as substrate to act upon them. This fact came into consideration by isolating cDNAs of these enzymes and further by enzymatic enactment of recombinant B5GT and B6GT. Gentian 3cGT and B5GT are present in the same cluster. On the other hand, the second route is supported by Sciuto and coworkers according to which in feather cockscomb (*C. cristata*), cyclo-DOPA 5-O-glucoside was an effective harbinger of amaranthine (betanidin 5-*O*-b-glucuronosylglucoside), as compared to betanidin or betanin. It is also strengthened by accumulation of cyclo-DOPA 5-O-glucoside in root peels of red beet and young beet plants. In addition, using phytochemical complementation, it has been shown that cyclo-DOPA accumulation occurs in mutant *M. jalapa* plant petals.

4.2.1.16 Conversion of Betanidin to Betanin

Conversion of betanidin to betanin is supported by betanidin-5-*O*-glucosyltransferase enzyme that connects the glucose moiety of uridine diphosphate glucose (UDP-G) and -OH group in the fifth position [10]. Formed betanin is transferred and stored to cytoplasmic vacuole later on [8].

Back transformation of betanin to betanidin is also possible due to the activity of β-glucosidase. Aside from the general accepted method of glucosylation of betanidin, betanidin is also synthesized by the action of 5-O-cyclo-DOPA glucosyltransferase that speeds up the transfer of sugar to cyclo-DOPA, and subsequently condensation occurs between the derived glucosides with betalamic acid. In *M. jalapa*, this activity is confirmed. In other betalain-synthesizing plants, preliminary tests have also been performed [6].

4.2.1.17 Conversion of Betanidin to Gomphrenin

Another glucosyltransferase, 6-*O*-glucosyltransferase (6-GT), is also identified, which performs catalysis of a corresponding reaction to 5-GT to produce gomphrenin I. From cell cultures of *D. bellidiformis*, both glucosyltransferases have been extracted. It seems conceivable that the reaction sequence may also depend on plant genus. Acylation of glucosylated betanidins takes place later on to produce 1-*O*-acylglucosides. This reaction is a specialty of betalain-synthesizing plants because it's analogous to acylation reaction of flavonoid which follows hydroxycinnamoyl-Co-A pathway. For example, in *C. rubrum*, an enzymatic activity of 1-*O*-hydroxycinnamoyl-transferase (HCA) is recognized that speeds up the relocation of hydroxycinnamic acids from 1-*O*-hydroxycinnamoyl-β-glucose to the C-2 hydroxy group of glucuronic acid of betanidin 5-*O*-glucuronosyl glucose (amaranthine) yielding celosianins [7].

4.2.1.18 Glucosylation of Cyclo-DOPA

Cyclo-DOPA is glucosylated by the catalytic activity of glucosyltransferase in the presence of UDP, thereby producing leuko-DOPA-chrome glucoside also known as cyclo-DOPA-glucoside.

4.2.1.19 Glucuronosylation of Glucosylated Cyclo-DOPA

After the formation of cyclo-DOPA or glucosylated cyclo-DOPA, betalains are further modified with sugar and acyl moieties rather than at betanidin (aglycone) stage. In recent years, the activity of UDP-glucuronic acid:cDOPA 5-glucoside glucuronosyltransferase is identified in *C. cristata*, which proves that the modification with the glucuronic acid moiety takes place at cyclo-DOPA stage.

As with the determination of activities of acyltransferases (which are acylglucose dependent) toward betanidin glucosides, acylation reactions in betacyanin biosynthetic pathway can be understood by isolating cDNAs of SCPL protein family which are then subjected to functional analysis. The glucosylation and acylation steps vary in different species, and both routes may be followed. Evidently, further discoveries are still needed [17].

4.2.1.20 Transformation of Betanin into Betanin Phenoxy Radical

Although glucosylation reactions diminish betanin lability for betanidin, the enzyme peroxidase is still capable to oxidize the remaining hydroxyl group. Betanin phenoxy radical is formed by betanin that is able to further the evolution resulting in the hydrolysis of molecule. Decarboxylated and dehydrogenated betacyanins are formed by rearrangement of oxidation products of betanin.

4.2.1.21 Formation of Decarboxylated Betalains

Decarboxylated betacyanins are identified in extracts obtained from *Celosia* sp. and *Carpobrotus acinaciformis* and in hairy root cultures of *B. vulgaris*. Unlike the presence of leuko-DOPA-chrome in the more common compounds, a leuko-dopamine-chrome moiety is present in these betacyanins. It is not identified that in biosynthetic pathway, decarboxylation takes place on which point and there's a possibility of two alternate routes [6]. Theoretically, decarboxylation may occur at C2, C15, and C17 of betacyanins and at C11 and C13 of betaxanthins [18]. Decarboxy-betacyanins may be synthesized from betaxanthins in much the same manner as described for the carboxylated pigments via tyrosinase-regulating reactions. Dopamine-betaxanthin (miraxanthin V) is the decarboxylated analog of DOPA-betaxanthin, whereas tyramine-betaxanthin (miraxanthin III) is that of tyrosine-betaxanthin. In *B. vulgaris* and *Celosia* sp., the co-occurrence of decarboxylated betacyanins and dopamine-betaxanthin supports this possibility. At L-DOPA or free tyrosine level, decarboxylation may take place, thereby forming dopamine or tyramine, respectively. Dopamine chrome can be formed by a spontaneous and tyrosinase-catalyzed reaction, and this may be transformed by the action of reducing agent to leuko-dopamine-chrome. Then condensation reaction of betalamic acid may produce decarboxylated betacyanins [6].

4.2.1.22 Net Level of DOPA

DOPA is consumed in three reactions as follows:

I. In the formation of betalamic acid.
II. In the synthesis of cyclo-DOPA and dopaxanthin.
III. In dopamine synthesis.

The dopamine synthesis ultimately synthesizes 2-descarboxy-betanidin. Consumption of DOPA may be increased in different pigment profiles of plants. Equilibrium between formation and consumption of DOPA may be established as indicated by the non-accumulation of their products in the same plant. Before and after process of betalain biosynthesis, net level of DOPA remains the same. An exceptional case is that of yellow flowering variety of genera *Glottiphyllum* and *Lampranthus* present in the family Aizoaceae which produces dopaxanthin as an exclusive betalain pigment. The maximum L-DOPA is consumed in betalamic acid and in cyclo-DOPA formation. This implies that cyclo-DOPA is not synthesized in yellow beet and parts of various plants of *Glottiphyllum*, *Lampranthus*, and *Portulaca* in which only betaxanthins are accumulated. CYP76AD1 expression is very small in these plants, which plays a role in the production of cyclo-DOPA; on the other hand, synthesis of betanidin may be catalyzed by the gene product DO/CYP76AD1 [5] (Figs. 4.1 and 4.2).

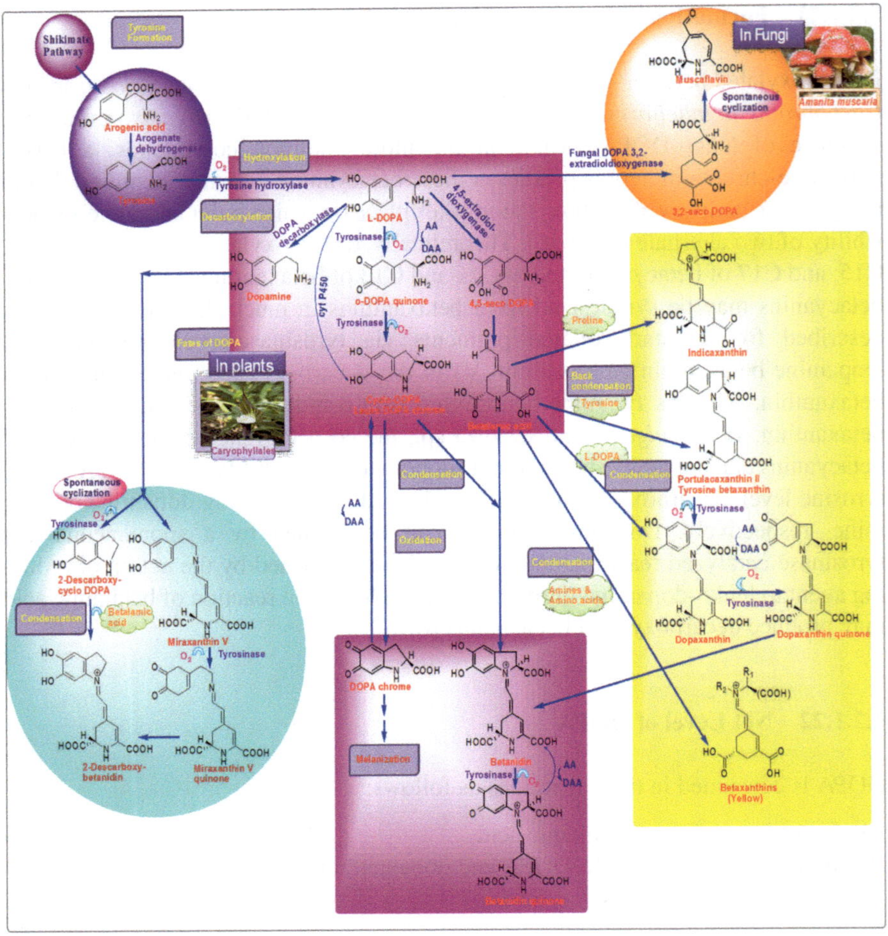

Fig. 4.1 Biosynthetic pathway of betalains using tyrosine as a starting molecule

4.2.2 Biosynthesis of Betalains from Tyramine

In this newly proposed biosynthetic pathway of betalains, tyramine serves as a starting material. This pathway was described. He proposed this pathway due to the facts that in the inflorescence of *Celosia* sp., hairy root cultures of yellow beet, and flowers of *Carpobrotus acinaciformis* of the family Aizoaceae, 2-descarboxy-betacyanin is found that is a subgroup of betacyanin pigments. The pathway comprises of two routes A and B which are detailed below.

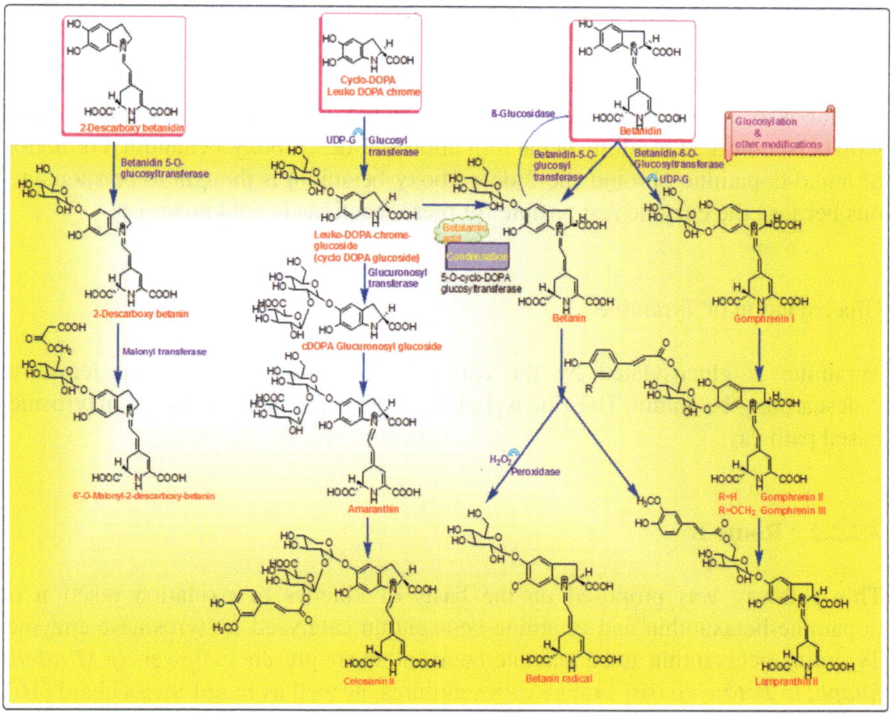

Fig. 4.2 Glucosylations and other modifications in tyrosine-based biosynthesis of betalains

4.2.2.1 Route A

This route comprises of the following steps.

Hydroxylation of Tyramine

Tyramine, the initial substrate of the pathway, is hydroxylated to dopamine first of all. This hydroxylation step is favored by tyrosine hydroxylase enzyme.

Oxidation of Dopamine

Dopamine produced as a result of hydroxylation is subjected to undergo oxidation by the polyphenol oxidase activity of tyrosinase, and dopamine-quinone is produced.

Condensation with Betalamic Acid

Dopamine quinone is transformed by an unknown enzyme to leuko-dopamino-chrome also known as 2-descarboxy-cyclo-DOPA which undergoes condensation with betalamic acid resulting in the formation of 2-descarboxy-betanidin. Formation of leuko-dopamino-chrome and 2-descarboxy-betanidin is thought to be spontaneous because the enzyme responsible for these reactions is not known yet.

Glucosylation of Tyramine

Tyramine is glucosylated by the activity of betanin 5-glucosyltransferase to 2-descarboxy-betanidin. The glucosylation occurs in the same manner as in tyrosine-based pathway.

4.2.2.2 Route B

This pathway was proposed on the basis of kinetics of oxidation reaction of dopamine-betaxanthin and tyramine-betaxanthin catalyzed by tyrosinase enzyme. Tyramine-betaxanthin and dopamine-betaxanthin are present in flowers of *Mirabilis jalapa*, in *Beta vulgaris*' orange callus cultures, as well as in red Swiss chard [16]. The following reactions occur in this biosynthetic route:

Hydroxylation of Tyramine

Hydroxylation occurs in the same manner as described in route A. Dopamine is produced by the hydroxylase activity of tyrosinase.

Condensation with Betalamic Acid

Dopamine undergoes condensation with betalamic acid directly in contrast to route A in which dopamine is first converted to 2-descarboxy-cyclo-DOPA. Condensation of betalamic acid and dopamine yields dopamine-betaxanthin.

Tyramine may also condense with betalamic acid directly (i.e., without the formation of dopamine) resulting in the formation of tyramine-betaxanthin which is then converted to dopamine-betaxanthin by hydroxylation catalyzed by tyrosinase hydroxylase.

Oxidation of Dopamine-Betaxanthin

Phenol oxidase activity of tyrosinase oxidizes dopamine-betaxanthin to dopamine-betaxanthin-quinone.

Intramolecular Cyclization

Dopamine-betaxanthin-quinone undergoes intramolecular cyclization as a result of which 2-descarboxy-betanidin is produced [8] (Fig. 4.3).

4.3 Biosynthesis of Betalains in Anthocyanin-Producing Plant Species

Biosynthesis of betalain in plants from tyrosine as a starting point omits the biosynthesis of anthocyanins (forerunner phenylalanine). This may be explained by the fact that in betalain-synthesizing plants, enzyme anthocyanidin synthase is not present which is essential for catalysis of the last step in biosynthetic route of anthocyanins [10].

No plant species is reported for the accumulation of both pigments: anthocyanins and betalains. The reason for the absence of their co-occurrence is still unknown. On the other hand, it's surprising that homologous gene for a key enzyme in the biogenesis of betalain is also present in anthocyanin-accumulating plants and the enzyme which glucosylates the betacyanins; betanidin 6-*O*-glucosyltransferase can also catalyze the glucosylation reactions of anthocyanidins.

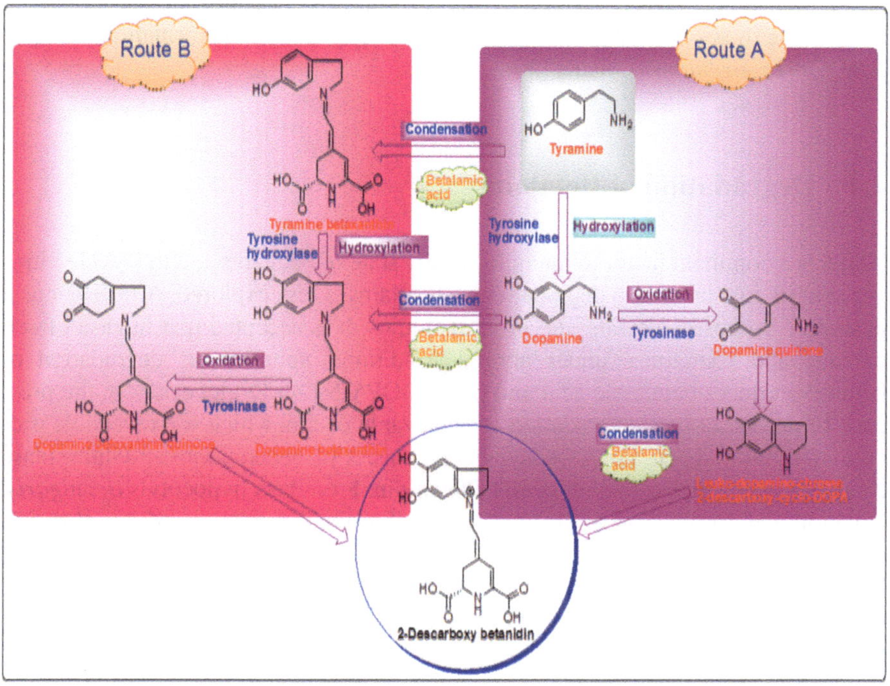

Fig. 4.3 Biosynthetic pathway of betalains using tyramine as a starting molecule

Furthermore there seems no apparent physiological obstructer that makes the co-occurrence of both betalains and anthocyanins impossible. Recently a part of the betalain biosynthetic route is transferred in potato cell cultures, *Antirrhinum* petals, and shoots of *Arabidopsis*, which were fed with L-DOPA, a betalain intermediate. This revealed the possibility of betalain synthesis in anthocyanin-producing plants.

Studies reveal that the ability of betalain synthesis may have arisen more than once and that some lineages in the course of evolution patently shift back from betalain production to anthocyanin production. So, a cost is associated with production of betalains, which prefers the evolution of anthocyanin-producing plants upon betalain-producing plants [19].

Production of both pigments betaxanthins and betacyanins can be done by introducing a single enzyme (DOD) and feeding its substrate (L-DOPA). In betalain synthesis, many steps occur spontaneously, so the betacyanin production is surprising. It suggests the background for the presence of an enzyme which performs the conversion of L-DOPA to cyclo-DOPA or dopaxanthin to betacyanin/betanidin and also the enzyme O-glucosyltransferase that acts against betanidin. To produce betaxanthin and betacyanin in anthocyanic plants, introduction of just two enzymes, DOD and another enzyme that converts tyrosine to L-DOPA (not identified yet), is sufficient. In Caryophyllales, initial evolution of this route described the new biosynthetic steps. In Caryophyllales and certain fungi, multiple origins of betalain biosynthetic pathway are supported by a relatively simple mechanism. By using the extensive study of anthocyanin biosynthesis and its function in *Arabidopsis*, the physiological effects were studied for the first time on the same plant tissues which accumulate both betalains and anthocyanins by stable transformation of 35S: AmDOD in *Arabidopsis* plants [20].

4.4 Degradation of Betalains

Albiet is a dominant factor to maintain steady state of betalains and to recycle nitrogen, but its role in in vivo degradation of betalains is less explored. A little about it is studied by postharvesting and in vitro studies. It is accepted that in case of cell damage and sometimes under normal conditions, spontaneous degradation of betanin produces betalamic acid and cyclo-DOPA-5-*O*-b-D-glucoside. In the presence of proper amines, degradation of betacyanin occurs which is a reversible process and may happen along the synthetic process of betaxanthins. β-Glucosidase catalyzes the degradation of betanin to betanidin. From beet hypocotyls *B. vulgaris*, isolation and purification of enzyme responsible for the catalysis of discoloration and breakdown of betacyanins were done. Oxidative discoloration of betanidin and betanin was induced by the activity of this enzyme. The enzyme may be a polyphenol oxidase (PPO); it degrades pigments in the presence of oxygen. It is membranous, and its activity is retarded by chelating agents. It is thought that peroxidases (EC1.11.1.7) may be involved in the degradation of betalains. Betanidin and betanin are oxidized by a protein fraction having peroxidase activity against hypocotyls of guaiacol from *B. vulgaris*. By oxidation of betanidin in the presence of enzymes, betanidin-quinone was formed as a sole product, whereas betanin oxidation yields betalamic acid and many other polymers of oxidized cyclo-DOPA-5-O-b-D-glucoside [4] (Fig. 4.4).

4.5 Production of Betalains by Stable Transformation of Arabidopsis

By 35S: AmDOD transformations mediated by *Agrobacterium*, stably transformed *Arabidopsis* plants were synthesized. Seedlings of T2 generation were examined for 35S: AmDOD transgene expression, and selected lines were subjected to experiments to check their betalain-synthesizing potential after feeding with L-DOPA. Within 12 h after feeding the whole seedlings with L-DOPA, new pigmentation was visible, also in hypocotyls and in etiolated root tissues. Pigments ranged from light yellow to orange and intense orange-red. When examined under blue light, green autofluorescence was produced by the pigment that is the characteristic of betaxanthins. When inflorescences isolated from highly grown plants were nourished with L-DOPA, after 24 h light yellow coloration appeared in whole petal tissues, stems, and siliqua, and strong autofluorescence also appeared. If L-DOPA is not fed in 35S: AmDOD seedlings or inflorescences, visible pigments are not formed, and when examined under blue light, significant autofluorescence was not produced. To study the nature of pigments produced by 35S: AmDOD *Arabidopsis* and following L-DOPA feeding, HPLC analysis was used. L-DOPA was fed in seedlings, and for chemical analysis, the entire seedlings were sampled. Several peaks present in tissues bombarded with 35S: AmDOD and fed with L-DOPA were revealed by HPLC analysis but not in the control tissue. Spectral data and retention times of four peaks, 10.4, 11.6, 20.4, and 21.7 min, did not show the presence of betaxanthins due to their smaller quantity. When seedlings were examined under blue light, strong autofluorescence showed the presence of betaxanthins. In spite of the orange-red coloration, betacyanin production was failed to be detected due to the ability of DOD to synthesize both betaxanthin and betacyanin when expressed transiently in petals of *Antirrhinum*. Orange-red color of seedlings

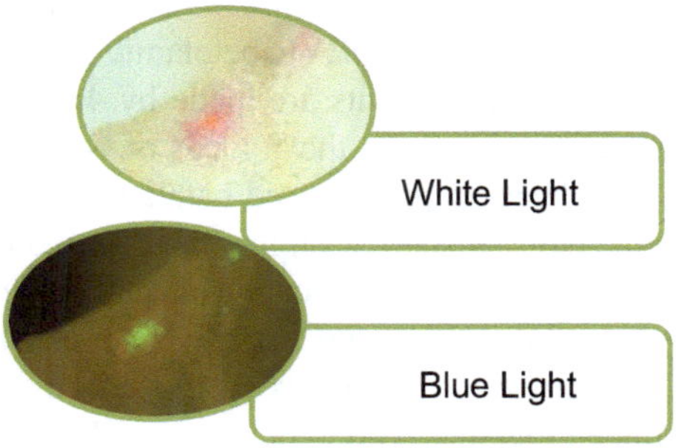

Fig. 4.4 Examination of pigment production by transient expression of 35S: AmDOD and feeding with L-DOPA under white and blue light

revealed the presence of betacyanins as proved by HPLC analysis, while the strong production of betaxanthins is indicated by autofluorescence. Moreover, red coloration did not appear for the nourished inflorescences of the adult trees. Thus it seems that betaxanthins are synthesized highly in the specific background coloration of seedlings due to which orange-red color appeared (Fig. 4.5).

(A to E) saplings were grown up on dampened sieve rounds with 10 mM L-DOPA and (F and G) without 10 mM L-DOPA. (A to C) were fed with L-DOPA, and they produced yellow to reddish orange colors that exhibited autofluorescence under the influence of blue light that is a characteristic of betaxanthins, as shown in hypocotyl (E) and cotyledon (D). The saplings that were not nourished with L-DOPA did not generate any color or pigment in their hypocotyls (G) or cotyledons (F) (Fig. 4.6).

4.6 Genetics of Betalains: Beets, Portulaca, and 4 O'Clocks

In the 1930s beets (*B. vulgaris*) were subjected to genetic analysis by Wesley Keller who identified two loci in betalain biosynthesis: R and Y. Red vs. yellow shift is due to R alleles that are related to the transformation of L-DOPA into cyclo-DOPA. It is observed that R locus encodes CYP76AD1. A dominant R allele is present in red table beets, whereas yellow beets have identical recessive r alleles at corresponding chromosomal loci. Whether (red or yellow) pigmentation occurs or not in the interior part of red beets is decided by Y alleles. Almost all sugar beets are included in white beets that have identical recessive y allele at corresponding chromosomal loci. In spite of the presence of recessive y allele, "white" beets still have the ability for betalain synthesis, and in this case the pigment is only located on epidermal layers. During cultivation, dominant Y allele was discovered that is responsible for red table beets. In Correns' work that was published in the early 1900s in Germany and was reviewed by Rheinberger, there was a classic work on *Mirabilis jalapa* (4 o'clock), and in a recent but still old work, it indicates two principal loci present in betalains, named as R and Y. Again, red vs. yellow pigmentation is due to the R alleles, and due to the R gene of beet, the yellow rr mutant gene is complemented to red. Whether the pigment is present or absent on the whole is decided by recessive y versus dominant Y alleles of 4 o'clock, respectively. In the biosynthetic pathway of betalain in beets as well as in 4 o'clock, the locus R is coherent with the production of cyclo-DOPA from L-DOPA. The Y locus of 4 o'clock is consistent with the conversion of L-DOPA to betalamic acid that is assisted by the DODA gene, because no betalain pigments are made by 4 o'clock yy mutants, while with few types of upstream regulator, the Y locus of beet is more consistent due to the strong ability of yy beets to produce red betalain pigments that are restricted only to the epidermis. Normal state of pigment is directed by the recessive y allele of beets that imitates the epidermal location of anthocyanins in a number of floral plants. Keller showed that the R and Y loci were connected at almost 7 cM in beets and this space is verified newly. Engels et al. showed that the R and Y loci of 4 o'clock are not linked together. Three loci C, R, and I direct the color of flowers in *Portulaca*.

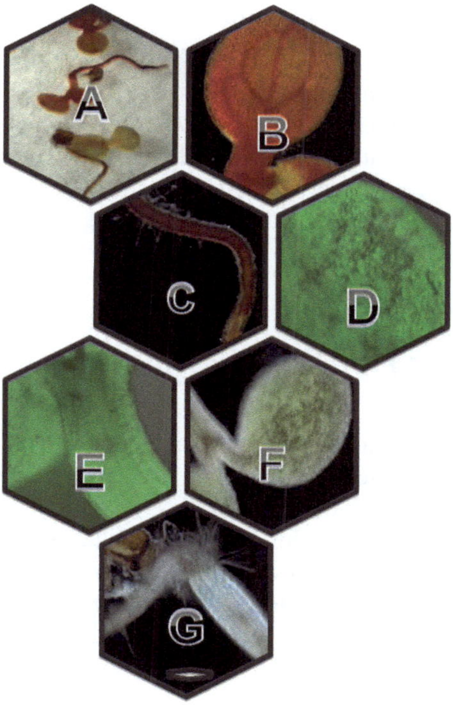

Fig. 4.5 Production of betaxanthin in *Arabidopsis* saplings firmly altered with 35S: AmDOD

Fig. 4.6 Pigmentation in *Arabidopsis* plant inflorescences transformed stably with 35S: AmDOD

4.7 Key Enzymes

Three important enzymes used in the biosynthesis of betalains include two oxidases tyrosinase and 4,5-DOPA-extradiol-dioxygenase that use molecular oxygen for their catalytic activity. The third enzyme betanidin glucosyltransferase transfers a sugar moiety and thus forms the structural unit of betacyanins. In plants, at transcription stage, the enzymes are governed, which are involved in the pathway [6] (Fig. 4.7).

4.7.1 Tyrosinase

Tyr is a plant PPO which contains copper nuclei and is classified under type 3 copper-centered proteins [5]. This enzyme is bifunctional; it acts as oxygenase and also as oxidase [5, 21].

Their function is to speed up the hydroxylation reaction of phenols, thus producing *o*-diphenols (monophenol/monooxygenase; EC 1.14.18.1) which generate *o*-quinones (*o*-diphenol/oxygen oxidoreductase; EC 1.10.3.1) upon oxidation [22].

4.7.2 Tyrosinase Source

Tyrosinases occur in fungi, bacteria, animals, and plants [8]. Fruits of *P. americana*, which are related to the assemblage of betacyanins, are characterized for the presence of copies of tyrosinase. In *S. salsa* and *B. vulgaris*, tyrosinase also takes part in pigment assemblage [6]. In *A. muscaria*, activity of tyrosinase has been identified which is related closely to the betalamic acid and muscaflavin biogenesis.

4.7.3 Tyrosinase Mode of Action

The tyrosinase does not act only on tyrosine; rather it shows low affinity and high Km value (0.3 mM) to other analogues; the protein has a molecular weight of 50,000 kDa and is thought to contain a heterodimeric structure. Its location is restricted to the colored part of the fungus where betalains and DOPA are assembled. PPOs of plants and fungi show dual function, i.e., they utilized DOPA as a reaction substrate as well as tend to synthesize it. The pure enzyme catalyzes the fabrication of DOPA from tyrosine and is accelerated by Fe^{+2} and Mn^{+2}; its activity is retarded by metal chelating agents [4].

Aside from browning reactions, PPOs of plant have to perform synthetic function of pigments such as betalains, nordihydroguaiaretic acid, and aurones which have low molecular weights. A type 3 protein having copper incorporation in the

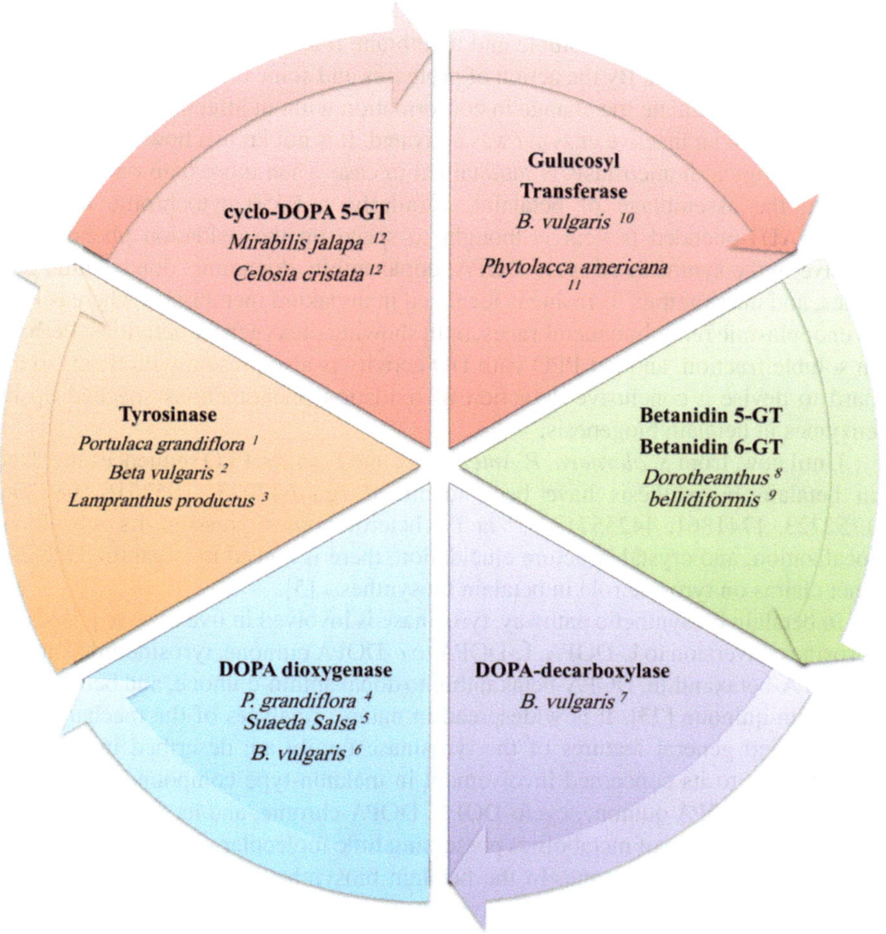

[1] Steiner, 1999, [2] Gandia Herrero et al., 2004, [3] Gandia Herrero et al., 2007, [4] Christine et al., 2004, [5] Ruan, 2008, [6] Zyrd et al., 2003, [7] Terradas, 1989, [8] Vogt et al., 1999, [9] Vogt, 2002, [10] Sepulveda-Jimenez et al., 2003, [11] Nakayama et al., 2008, [12] Sasaki et al., 2005

Fig. 4.7 Enzymes involved in biosynthesis of betalains

center is present in the membrane of thylakoid; a PPO known as catechol oxidase acts on a little of L-tyrosine substrate. Its function is not discovered yet, and its structure differs from tyrosine in having an additional C-terminal region. The plant PPO gene involved in betalain biogenesis was found first in *P. americana* and has a deduced order of amino acids which consist of a plastid transit peptide. An enzyme was purified by taking into account the above information of gene sequence and further reports on hydroxylation of tyrosine in *P. grandiflora* by performing inhibition experiments with Cu^{2+}-chelating agents. It has dual function which is indicated by two catalytic activities shown by this: TOH (EC1.14.18.1) and DO (EC 1.10.3.1).

It was thought earlier that more than one PPO are involved in the biogenesis of betalains. Moreover, from soluble and membrane fraction of red beetroot, an inactive PPO was purified. By the action of proteases and some other agents which have the ability to stimulate the change in conformation without affecting the active site, DO activity of an inactive enzyme was activated. It is not known how the regulation of physiological homeostasis is maintained in case when more than one PPO take part in the assemblage of betalains. Similarly, a P450 cytochrome that is a CYP76AD1-encoded protein is thought to speed up the oxidation phenomenon involve in the synthesis of cyclo-DOPA, dopaxanthin, dopamine, dopaxanthin quinines, and miraxanthin. Tyrosine is localized in thylakoid membranes whereas PPO in endoplasmic reticulum membranes, both showing dioxygenase activities. Perhaps in soluble fraction, another PPO with DO activity is also present with them, so it's hard to devise a conclusive depiction of oxidation phenomenons speeded up by enzymes in betalain biogenesis.

Until now, from *S. oleracea*, *P. Americana*, and *S. salsa*, PPO sequences involved in betalain biosynthesis have been identified (gi: 642023, 1052516, 984206, 1752723, 1741861, 442557134). Via Tyr heterologous expression, its subcellular localization, and crystal structure elucidation, there is a need to streamline the distinct claims on tyrosine role in betalain biosynthesis [5].

In betalain biosynthetic pathway, tyrosinase is involved in five diverse phases: in tyrosine conversion to L-DOPA, L-DOPA to *o*-DOPA quinone, tyrosine-betaxanthin to DOPA-betaxanthin, DOPA-betaxanthin to dopaxanthin-quinone, and betanidin to betanidin-quinone [23]. It is widespread in nature. Analyses of the mechanism of enzyme and general features of the tyrosinase family are described in literature widely. Due to its concerned involvement in melanin-type compound production, L-DOPA, o-DOPA quinone, cyclo-DOPA, DOPA-chrome, and its melanization, its part in the synthesis of metabolites possessing little molecular weight was not taken into account for a long time. In the betalain biosynthetic route, disregarding the nature of reactions catalyzed by tyrosinase, the tyrosinase activity could be extremely controlled to ensure the pigment production in accurate tissues at precise time.

From plants capable to synthesize betalains, a sequence signature symptomatic for tyrosinases is not identified yet unlike the enzyme 4,5-DOPA-extradiol-dioxygenases. Perhaps, it is due to a little of tyrosinase orders accessible from Caryophyllales such as *P. americana* and spinach (*Spinacia oleracea*) (gi: 642023, 984206, 1052516, 1741861, 1752723). In conformity of the results found for the tyrosinase involved in aurone biogenesis, an appropriate atmosphere due to the enzyme that forms betanidin by cyclization of dopaxanthin-quinone has been devised. In betalain biosynthetic pathway, the tyrosinase-assigned reactions may be catalyzed by several other enzymes, as cytochrome P450 is associated in the L-DOPA transformation into cyclo-DOPA by gene silencing and genetic complementation in *B. vulgaris*. In a new cytochrome P450 subfamily, tyrosinase (gi: 356968415) is the first member with homologues recognized in other plants synthesizing betalain such as *M. jalapa* (gi: 356968421) and *A. cruentus* (gi: 356968419) [6].

4.7.4 4,5-DOPA-Extradiol-Dioxygenase

From *A. muscaria*, an enzyme convoluted in betalain biosynthesis, L-DOPA dioxygenase (DODA) was purified and characterized. It is bifunctional; L-DOPA-4,5-dioxygenase forms betalamic acid, whereas L-DOPA-2,3-dioxygenase catalyzes the muscaflavin biosynthesis [5].

The ferrous (Fe^{+2}) and ferric (Fe^{+3}) ions are an essential cofactor in these enzymes. From *A. muscaria*, Girod et al. isolated the DOPA 4,5-dioxygenase enzyme that is an oligomer like other extradiol-cleaving dioxygenase enzyme that catalyzes the 4,5-extradiol interruption of L-DOPA causing the formation of betalamic acid via 4,5-seco-DOPA intermediate. By affinity chromatography, it is revealed that this DOPA enzyme consists of a variable amount of similar 22 kDa fractional monetary units. From *A. muscaria*, extraction of one more enzyme, DOPA 3,4-dioxygenase, was carried out which speeds up the 2,3-ring cleavage, thereby producing a pigment muscaflavin that is such compound that is not ever present in plants [2].

4.7.5 Non-heme Dioxygenases

Non-hemeproteins consisting of iron such as dioxygenases speed up the ring opening of catechol compounds by incorporation of oxygen molecule [24]. From *P. grandiflora*, the first experimental evidence of 4,5-DOPA-extradiol-dioxygenase was characterized. In bacteria, from *M. jalapa*, firstly the expression of recombinant protein was obtained for the 4,5-DOPA-dioxygenase, which proved in vitro the betalamic acid production. From *B. vulgaris*, the first enzyme capable to produce betalamic acid was purified. For flowering plants of *M. jalapa*, working of betalain biosynthetic 4,5-dioxygenase was investigated. At the final stages of developing colored flowers, enzyme copies were identified but not in other plant parts. Enzyme obtained from *A. muscaria* was subjected to Northern blot analysis which demonstrates the transcriptional level regulation, with mRNA assembled mostly in the cap's colored tissues. 4,5-*Seco*-DOPA undergoes intramolecular cyclization to form betalamic acid which is a spontaneous reaction. However, the seco-DOPAs are present naturally without further evolution [6]. Moreover, in the betalain biosynthesis, only the involvement of 4,5-DOPA-dioxygenases of Caryophyllales plants is obtained. It is concluded if appropriate atmosphere for the condensation of intermediate is present; the formation of betalamic acid is favored by the righteous enzymes. By sequence analysis, it is clear that 4,5-DOPA-dioxygenases participated in the production of betalains comprise of a group of preserved amino acids near the active site unlike additional dioxygenases of plants [25]. This is advantageous in betalamic acid condensation reaction. It is revealed by phylogenetic examination of presently arranged dioxygenases that the dioxygenases that are present in betalain-producing plants constitute a group aside from early homologues of plants [6].

4.7.6 Glucosyltransferases

From betalain-synthesizing species, two groups of 5-*O*-glucosyltransferases were characterized. One group glycosylates betacyanin and betanidin; betanidin glucosyltransferase and the second group known as cyclo-DOPA glucoside perform the modification of cyclo-DOPA. It seems that both scenarios are occurring at the same instance. Betanidin-glycosylating enzymes are of two types, betanidin 5-GT (5-*O*-glucosyltransferase) and betanidin 6-GT (6-*O*-glucosyltransferase). Transfer of glucose to betanidin 5 hydroxyl group from UDP-glucose takes place by betanidin 5-GT. To the 6-hydroxyl group, the glucose transfer occurs by betanidin 6-GT.

Betanidin 5-GT and betanidin 6-GT are non-particular, and glucose transfer to various anthocyanidins and flavonols can be catalyzed by them that depend on activities and sequence of protein. It is assumed that flavonoid glucosyltransferases evolve these enzymes [12]. Identity of only 15% amino acid sequence is shown by the two betanidin glucosyltransferases, which is very little than the expected sequence in the case when both enzymes share a common ancestor [22].

The transmission of sugar molecules to particular acceptors, including pigments as hormones, secondary metabolites, and ecological poisonous complexes, is catalyzed by glucosyltransferases. The glucose addition to betanidin is regulated regioselectively at any of the two OHs in order to produce gomphrenin I or betanin. From *D. bellidiformis*, betanidin glucosyltransferases (gi: 5918023, 18033791) were first made pure and characterized, and by location-directed mutagenesis and homology modeling, the characterization of the active site was made which explained the betanidin transformation into betanin successfully. No grounds for betaxanthin glycosylation have been reported in vivo or in vitro despite its structural analogies with betacyanins. In the literature, there are 31 glucosyltransferase enzyme sequences present in betalain-synthesizing plants that are set down up to now. In the pathway, only a few sequences are recognized to be active toward substrates. From *B. vulgaris*, a glucosyltransferase (gi: 29692096) is involved in the betanin production, while other sequences present in the same plant (gi: 46430997, 46430995) glucosylate betanidin to a lesser magnitude as compared to other substrates. In *M. jalapa*, a glucosyltransferase is characterized that acts on cyclo-DOPA instead of on betanidin (gi: 62086401). For *Celosia cristata* glucosyltransferase (gi: 62086403), the same activity is identified. In *B. vulgaris*, regulation of glucosylation is indicated at the level of transcription. Activity of glucosyltransferase and changes produced in the amount of enzyme copies are correlated positively with the accumulation of betanin in *M. jalapa*'s developing flowers [6].

4.7.7 Hydroxycinnamoyl Transferases

Betanidin glucoside acylation (EVI) with hydroxycinnamic acids takes place through the catalysis of hydroxycinnamoyl glucoses (β-acetal esters) that donate acyl groups as found in the eight plant species of four diverse families present in

Caryophyllales order. This is strengthened by the fact that respective 1-O-acylglucosides are always present in hydroxycinnamoyl betacyanins. In acylation of betacyanin, the acceptance of hydroxycinnamoyl-CoA thioesters which are alternative acyl donors is not confirmed. It is contrary to the studies of common hydroxycinnamoyl transfer reactions which involve thioester in the biosynthesis of flavonoid. In the biosynthesis of flavonoids, the substrate may be β-acetal esters as indicated in *Daucus carota* by the formation of cyaniding hydroxycinnamoyl-triglycosides. It is revealed that acyl transferases that depend upon 1-O-acylglucoside are developed from serine carboxypeptidases as given in some recent papers. This was claimed for the dibutyryl glucose formation in *Lycopersicon pennellii* and sinapoyl malate formation in *Arabidopsis thaliana*. So, cloning few hydroxycinnamoyl transferases involved in betalain production is an interesting job which may be related possibly to proteins resembling serine carboxypeptidase.

4.7.8 DOPA Decarboxylase

Betaxanthins and betacyanins originated from DOPA are produced by the decarboxylating enzymes that decarboxylate either tyrosine (with consequent hydroxylation) or DOPA. Terradas in 1989 detected an enzyme that catalyzes DOPA decarboxylation to dopamine, in protein extracts from red beet cell cultures. On crude protein extracts obtained from yellow beet's hairy root cultures, similar experimentation was done that revealed DOPA to dopamine conversion taking place by DOPA decarboxylase because tyrosine was not accepted as a substrate [22].

4.8 Biosynthesis of Betalains in Potato Cell Cultures Via DOD Transient Expression

By 35S particle bombardment, GFP (green fluorescent protein) or *A. muscaria* DOD (35S: AmDOD) or *P. grandiflora* DOD cDNA (35S: PgDOD) has constructs regulated by CaMV35S promoter. The suspension cultures of potato cell were transmuted and studied for synthesis of betalains later on by feeding with L-DOPA. The cells that were transmuted by 35S gave GFP detection after 24 h of biolistic transformation, but there appeared no pigment formation. In cells transmuted by 35S: PgDOD or 35S: AmDOD and later feeding with L-DOPA, there appeared multicellular bunches of pigments within a day of post-bombardment. Orange and yellow cell clusters were obtained from both 35S: AmDOD and 35S: PgDOD. Red pigmented cells were also formed by 35S: AmDOD. Red pigment production refers to the formation of betacyanins, whereas yellow pigmentation suggests betaxanthin production in the cells. Yellow-pigmented areas were observed under blue light that revealed autofluorescence properties of pigment betaxanthin.

4.9 Biosynthesis of Betalain in Antirrhinum Petals by DOD Transient Expression

By 35S particle bombardment, PgDOD, the ventral surface of *Antirrhinum* dorsal petals, was transmuted, and betalain synthesis was studied by means of or deprived of petal penetration with L-DOPA. Petals were transformed biolistically with 35S GFP vector, and additional control was provided with L-DOPA feeding. In flavonoid production, mutant *Antirrhinum* lines were casted off to make petal backgrounds that are free of anthocyanin and in which any pigmentation can be observed easily. In 35S transformed petals: PgDOD but not penetrated with L-DOPA, and in 35S: GFP shot tissue, there was no visible pigmentation although there were positive apparent GFP foci. The 35S PgDOD petals incorporated with L-DOPA contained many multicellular yellow-colored foci after 24 h of penetration (after 48 h of bombardment), which indicated the synthesis of betaxanthin. Under blue light, the yellow foci showed robust green self-induced fluorescence that is characteristic of betaxanthins.

Yellow or orange and pink or red multicellular foci appeared by bombarding the 35S: AmDOD construct, though the pink or red foci occurred only contradictorily. Likewise, in some repeated experiments with 35S: PgDOD, pink foci were present occasionally. The pink coloration is an indication of betacyanin synthesis. Central cells gave strong autofluorescence, but the pink regions did not, which indicated that betacyanins and betaxanthins were possibly accumulated around the central region. Yellow pigments that were produced by bombardment with 35S: PgDOD were analyzed by using HPLC technique (LC-DAD). HPLC analysis was performed with ridged region of petals, to ensure the sampling of the same region in each case. Betaxanthin absorbance maximum was around 470 nm, while that of betacyanins was around 538 nm, and the examination of peak profiles was done at these two wavelengths. By bombarding with 35S: PgDOD petals or 35S: GFP petal tissue without incorporation of L-DOPA, no peak related to betalain was detected with HPLC. In case when these were incorporated with L-DOPA, separate peaks were observed at 470 as well as 538 nm. By comparing the data of HPLC spectrum and their retention times against the standard extracted from beetroot and from the standard reported spectral data, the compounds were identified. Peak 1 was the expected peak of betaxanthin vulgaxanthin I obtained from the extract of beetroot; it has a retention time of 7.33 min with maximum wavelength of 259,469 nm; Peak 2 was for betacyanin betanin with a retention time of 21.37 min and maximum wavelength of 269,534 nm, while Peak 3 was of isobetanin with a retention time of 24.78 min and maximum wavelength of 269,534 nm. Samples mixed with 35S: PgDOD L-DOPA gave a minor peak of betanin (Peak 2), which revealed that the small amount of betanin was produced and an unknown peak (Peak 5) was also obtained with a retention time of 21.02 min and maximum wavelength of 241,470 nm that was of dopaxanthin most probably.

Generally, the same pattern of chromatograms at 538 nm as well as at 470 nm was obtained except smaller detection of vulgaxanthin I (Peak 1) and the supposed peak of peak (Peak 5), from which one more clue was obtained that Peak 5 was a betaxanthin. Yellow pigments can be produced naturally in the petal expression and gullet in *Antirrhinum*. These pigments are formed from flavonoid pathway and are called aurones. Spectral data with maximum range of 390–430 nm proved the formation of aurones by bombarding with 35S: PgDOD followed by L-DOPA feeding. Normally, aurones are absent in abaxial petals of used line of *Antirrhinum* (JI19). Although red pigmentation was not apparent, a small quantity of betacyanin was associated with *Antirrhinum* samples bombarded with the 35S: PgDOD followed by HPLC analysis. In potato cells that express 35S: AmDOD, it is believed that betacyanin is present, given the red-colored pigments of some foci.

It is not clear yet how the formation of betacyanin can exist in non-betalainic species by the introduction of DOD that acts on incorporated L-DOPA. Betalamic acid is produced by the activity of DOD on L-DOPA. In plant cells, due to the frequent ability of tyrosinases, an endogenous enzyme may be present that is capable to act on L-DOPA in many non-betalainic species of plants. On the other hand, by endogenous activity of tyrosinases on dopaxanthins, betacyanin may be synthesized. An *O*-glycosylated betacyanin is the most common betacyanin detected in *Antirrhinum* petals. This reveals that new betacyanin substrates may be catalyzed by endogenous glycosyltransferases. By action upon betanidin, two *O*-glucosyltransferases have been identified both having the same sequence like the *O*-glucosyltransferases involved in the biogenesis of anthocyanin, and in fact, the betanidin5-*O*-glucosyltransferase can act upon both flavonoids and betacyanins. Thus, it may happen that the *Antirrhinum*'s endogenous flavonoids O-glucosyltransferases can catalyze betanidin (and/or cyclo-DOPA). Bombardment with either 35S: PgDOD or 35S: AmDOD to produce biolistic transformation in both potato- and *Antirrhinum*-produced multicelled foci resulted in betalain synthesis. It would not happen as a result of DOD enzyme's movement between cells as in the case of GFP, because it is huge enough for intercellular movement. Assemblage of betalain pigments in vacuoles reveals the movement of some precursors in the cell. It was proved by the research of Mueller and his coworkers in which they incorporated AmDOD biolistically into various mutant *P. grandiflora* backgrounds. After bombardment, solitary colored cells were shown within 18 h which established the multicellular foci within 2 days after bombardment. It was proposed that betalains exhibit the ability of diffusion through plasmodesmata to nearby existing cells. When complementation of DOD mutant of *P. grandiflora* was directed with PgDOD, only a solo cellular foci was obtained, which is in contrast to the above results. Moreover, betalain synthesis is commonly cell-specific such as in petals' epidermal cells as observed in plants. As seen in our results, it may happen that high level of produced L-DOPA from tissue nourishing permitted the migration of betalamic acid not to the vacuole and closely existing cells.

4.10 Production of Betalains by Stable Transformation of Arabidopsis

By 35S: AmDOD transformations mediated by *Agrobacterium*, stably transformed *Arabidopsis* plants were synthesized. Seedlings of T2 generation were examined for 35S: AmDOD transgene expression, and selected lines were subjected to experiments to check their betalain-synthesizing potential after feeding with

L-DOPA. Within 12 h after feeding the whole seedlings with L-DOPA, new pigmentation was visible, also in hypocotyls and in etiolated root tissues. Pigments ranged from light yellow to orange and intense orange-red. When examined under blue light, green autofluorescence was produced by the pigment that is the characteristic of betaxanthins. When inflorescences isolated from highly grown plants were nourished with L-DOPA, after 24 h light yellow coloration appeared in whole petal tissues, stems, and siliqua, and strong autofluorescence also appeared. If L-DOPA is not fed in 35S: AmDOD seedlings or inflorescences, visible pigments are not formed, and when examined under blue light, significant autofluorescence was not produced. To study the nature of pigments produced by 35S: AmDOD *Arabidopsis* and following L-DOPA feeding, HPLC analysis was used. L-DOPA was fed in seedlings, and for chemical analysis, the entire seedlings were sampled. Several peaks present in tissues bombarded with 35S: AmDOD and fed with L-DOPA were revealed by HPLC analysis but not in the control tissue. Spectral data and retention times of four peaks, 10.4, 11.6, 20.4, and 21.7 min, did not show the presence of betaxanthins due to their smaller quantity. When seedlings were examined under blue light, strong autofluorescence showed the presence of betaxanthins. In spite of the orange-red coloration, betacyanin production was failed to be detected due to the ability of DOD to synthesize both betaxanthin and betacyanin when expressed transiently in petals of *Antirrhinum.* Orange-red color of seedlings revealed the presence of betacyanins as proved by HPLC analysis, while the strong production of betaxanthins is indicated by autofluorescence. Moreover, red coloration did not appear for the nourished inflorescences of the adult trees. Thus it seems that betaxanthins are synthesized highly in the specific background coloration of seedlings due to which orange-red color appeared.

(A to E) saplings were grown up on dampened sieve rounds with 10 mM L-DOPA and (F and G) without 10 mM L-DOPA. (A to C) were fed with L-DOPA, and they produced yellow to reddish orange colors that exhibited autofluorescence under the influence of blue light that is a characteristic of betaxanthins, as shown in hypocotyl (E) and cotyledon (D). The saplings that were not nourished with L-DOPA did not generate any color or pigment in their hypocotyls (G) or cotyledons (F) (Fig. 4.8).

4.11 Cloning the Betalamic Acid Biosynthetic Gene

4.11.1 4,5-DOPA-Dioxygenase

DODA biosynthetic gene responsible for ring structure obtained from *A. muscaria* mushroom was cloned. From mushroom caps, Zryd and coworkers purified a protein, and they developed antibodies against that protein and that anti-DODO antibodies were utilized for screening of *Amanita* cDNA expression libraries in *Escherichia coli* in order to clone DODA gene. They observed that mushroom DODA gene expression was complemented with white *Portulaca* tissue (cc genotype), enabling the red or yellow pigment production, depending on the R/r locus genotype. They also successfully cloned the first plant betalain

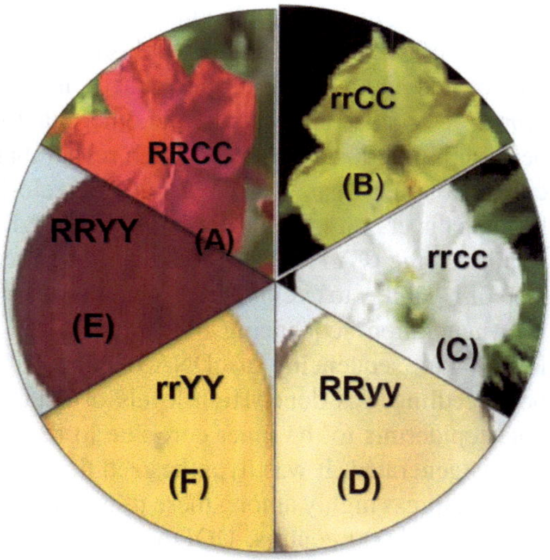

Fig. 4.8 Genetic phenotypes of betalains A,B,C 4 o'clock and D,E,F table beets

4,5-DOPA-dioxygenase (DODA. Unluckily, similarity between the mushroom gene and protein to the plant gene), and protein was not sufficient to allow the use of the antibody or the mushroom sequence of gene for further research. The subtractive libraries were generated that helped in the differential isolation of expressed cDNAs and identified genes highly expressed in red flowers of *Portulaca*. By sequences of clones, an acknowledged product of translation was identified with close resemblance to the Lig B ring cleavage world of proteins present in prokaryotes having capability of extradiol 4,5-dioxygenase activity like 4,5-DOPA-dioxygenase activity of betalains. It was came to know that this gene expressed itself in tissues of white *Portulaca* petals producing thereby yellow or red pigmentation, which depends upon their genetic background, revealing the presence of the right gene. The gene was subjected to sequence analysis that showed its different origin in phylogeny than the gene present in mushroom. It was pointed out that the presence of homologues of this gene in all plants and that the identification of DODA genes in betalain- and non-betalain-synthesizing plants are made by a common set of amino acid that changes near the putative active site. It is not concluded yet whether changes in specific substrate or same ancestors of Caryophyllales change these amino acids.

4.11.2 Cloning the Cyclo-DOPA Biosynthetic Gene, CYP76AD1

Conversion of tyrosine to L-DOPA and L-DOPA to cyclo-DOPA was assumed to be catalyzed by same enzyme, a PPO. The assumption was strengthened by the facts that tyrosinase, a PPO enzyme, catalyzes the conversion of both substrates in animals during the production of dark melanin pigments and PPO enzymes of plants catalyze these reactions in in vitro synthesis. PPO against debate is that biosynthetic pathway of betalain occurs in cytoplasm, whereas all known PPO enzymes of plants are located in chloroplasts and that it is a commonplace in 4 o'clocks, beets, *Portulaca*, cockscomb, cactus, and other plants to obtain mutants that produce only betaxanthin yellow pigments, which reveal that these mutants have an undamaged conversion of tyrosine to DOPA but the cyclo-DOPA ring structure is not available by the conversion of L-DOPA to produce red pigmentation. In beets, cytochrome P450 enzyme (CYP76AD1) was identified recently that is involved in the conversion of L-DOPA to cyclo-DOPA. Cytochrome P450 was identified by analyzing the next-generation sequencing of copies of red beet. By using pyrosequencing of Roche 454, sequencing of cDNA obtained from hypocotyls sections of red table beet seedling was done. High levels of betalain pigments are produced from the outer epidermis to the inner core due to these stem sections; thereby swollen red beet is generated. It was hypothesized that genes of betalain biosynthetic pathway would express highly among these transcripts. It was justified because in the data set of nearly 10,000 contigs, DODA gene of biosynthesis of betalains was the 14th most expressed. To check the involvement of other highly expressed genes in the formation of DOPA and its cyclization to cyclo-DOPA, the 454 database was studied. Laccase-encoding cDNAs and PPO were found, but they were not highly expressed like the genes involved in betalain biosynthetic. Query of contig database was done for sequences that encode cytochrome P450 because conversion of tyrosine to DOPA seems likely to the canonical cytochrome P450 reaction and conversion of DOPA to cyclo-DOPA was possible by the action of cytochrome P450. The 33rd contig that was expressed most highly was recognized as cytochrome P450 cDNA, CYP76AD1.

In different Caryophyllales that have the ability to synthesize betalains, similar transcript profiling was done, which revealed that CYP76AD1 is a founding member of a new subfamily of cytochrome P450 enzymes that resembles mostly to CYP76T and CYP76C. In 4 o'clocks, CYP76AD3 catalyzes the abovementioned step, whereas in *Amaranthus*, it is catalyzed by CYP76AD2. By various biochemical experiments and genetics, activity of CYP76AD1's role in the betalain biosynthesis was confirmed. CYP76AD1 is responsible for red phenotypes because in red beets, its high concentrations are present, whereas in white and yellow beets, it is expressed on low level. Through virus-induced gene silencing (VIGS), mutant analysis was done to lower the expression of this gene in very red beets that caused lowering the expression of red pigment, and yellow pigmentation became apparent. Via mass spectrometry, shift of pigmentation from red to yellow, i.e., from betanin to betaxanthin, was proved. Several yellow mutants are present, and CYP76AD1 complements the yellow mutants in cockscomb, beets, and 4 o'clocks. In all these plants, yellow phenotype is lost, and red pigments are produced due to overexpression of CYP76AD1. These data revealed that cyclo-DOPA formation by L-DOPA is catalyzed by CYP76AD1. If L-DOPA is not cyclized spontaneously, then red betalain pigments are not synthesized, but production of yellow pigments is not affected that is due to the function of the R gene present in beets. By these data, it was suggested that CYP76AD1 was the beet R gene that was defined genetically. In a variety of sugar beets, by sequencing CYP76AD1 alleles, C869, segregation of red (R) and yellow (r) hypocotyls was performed. Complete linkage to red phenotype (R) is revealed by the identification of a 5 bp insertion, 325 bp before the stop codon in the yellow segregate having rr phenotype (TAAAT). This insertion and the resulting frame shift introduced an early stop codon, which causes the deletion of the heme-ring-binding site and results in an inactive protein. More than 70 years ago, CYP76AD1 was identified as R locus on the basis of genetic data and the functional data which is described above.

4.12 Expression in Heterologous Species

Betalain pigments are synthesized only in Caryophyllales and certain fungi, but experiments are being done which are based only on DODA gene to introduce biosynthesis of betalains to some heterologous species such as yeast, *Antirrhinum majus* (snapdragon), *E. coli*, *Arabidopsis thaliana*, and potato (*Solanum tuberosum*). With genes of betalain biosynthesis, the first experiment was performed on *E. coli* that was unsuccessful. The first success to produce betalamic acid by using an *E. coli*-expressed DODA gene of 4 o'clock (Mj DODA) and its substrate L-DOPA was obtained by Sasaki and coworkers. They explained the expression of DODA gene in yeast by some preliminary experiments; but they did not give any experimental data. Recently, A DODA gene from *B. vulgaris* is expressed by Gandia-Herrero and Garcia-Carmona with its codons optimized for *E. coli*. Betaxanthin (yellow) pigments as well as betalamic acid were also synthesized successfully by them. They used the method of expressing the DODA gene in a heterologous system, and in vitro pigment production is carried out by using protein extracts. First of all, two different groups produced betalains in vivo in heterologous system. Harris and his coworkers expressed a DODA gene in petals of *A. majus* (snapdragon) and cell cultures of *S. tuberosum* (potato) from *Portulaca grandiflora* (PgDODA) by biolistically introducing the overexpression constructs. By using *Agrobacterium* transformation, they successfully produced stable transgenic lines of *A. thaliana* by expression of DODA gene PgDODA. When transgenic tissues were fed with L-DOPA, in vivo synthesis of red, yellow, and orange betalain pigments was carried out. A surprising thing is that by DODA gene expression and feeding with L-DOPA, yellow and red betalain pigments can be produced in non-betalain plant species. This suggests that L-DOPA can undergo cyclization to form cyclo-DOPA by a single activity in anthocyanic plants. But, it is still unknown how this activity was performed.

Hatlestad et al. also succeed in producing in vivo heterologous expression in yeast, with DODA gene and L-DOPA cyclization enzyme CYP76AD1. In yeasts, beet DODA BvDODA1 was expressed followed by L-DOPA feeding by them. They noticed the synthesis of yellow betalain pigments, betaxanthins. In yeast CYP76AD1 was expressed by itself in order to test its enzymatic activities. It was noted that production of cyclo-DOPA by CYP76AD1 uses L-DOPA as a substrate as in betalain synthesis. Also, in vivo synthetic pathway of betalains was recreated in yeast by expression of CYP76AD1 and BvDODA1 genes at the same time followed by L-DOPA feeding which resulted in the betanidin synthesis, an undecorated pigment of red beet [12].

4.13 Biosynthesis Regulation

Biosynthesis of betalains is a complex process due to the involvement of different environmental as well as physiological factors like the synthesis of other secondary metabolites [7]. Betalain pigments are assembled only in some tissues and at specific developmental stages. Light and cytokinins regulate betalain biosynthesis. Light, DOPA, and exposure to kinetin affect differently between different betacyanin- or betaxanthin-synthesizing plants [2]. For betalain regulation, auxin (2,4-D)-to-cytokinin (6-BAP) ratio is an important factor in *B. vulgaris* cell cultures [4].

By the influence of DOPA, light, or kinetin, betanin synthesis was induced. However, by feeding the seedlings with DOPA, betaxanthins were synthesized, and by supplying kinetin and light exposure, production of betalains was increased notably. If betalamic acid is present in plants, betaxanthins or both betacyanins and betaxanthins are produced, and if it is not present, only betacyanins are produced. Thus, the presence or absence of betalamic acid shows a regulatory mechanism in betalain biosynthesis. In plants, the mechanism of condensation of betalamic acid with cyclo-DOPA or glycolized-cyclo-DOPA has been suggested to generate betacyanins, thus stopping the betalamic acid accumulation. In contrast, to allow the betalamic acids assemblage in betaxanthin-synthesizing plants, control mechanisms must be different. In biosynthesis of betalains, regulatory mechanisms are not known well, but it has been suggested that hormone control and photo regulation are involved.

Due to the deficiency of DNA synthesis or slowdown of cell cycle, betacyanins assemblage was induced with suspension cultures of *P. americana*, by the inhibitors having different action modes of cell division. It was proposed that biosynthesis and accumulation of betacyanins are correlated with DNA synthesis as well as with cell cycle [2].

In red beet cell cultures, biosynthesis of betalains is regulated mainly by DNA methylation or rearrangement of genes as indicated by preliminary observations. Cytokinin (6-benzylamino purine; 6-BAP) and 2,4-dichlorophenoxyacetic acid (2,4-D) synthetic hormones of plants effect these changes. On the other hand, biosynthesis of betalains is inhibited when exogenous implement of natural hormone indole-3-acetic acid (IAA) is made. IAA oxidase, an enzyme responsible for the biosynthesis of indole acetic acid, is inhibited by betalains, and IAA inhibitory effect on elongation of wheat root is also weakened by betalains. In plants, IAA and betanin may compete to interact with a common signal transduction system, which can accept betanin more due to the presence of substructure indole-2-carboxylic acid. This reveals that IAA may mimic betalains and cause feedback suppression of betalain production. On tyrosine IAA, as well as 2,4-D, has a regulatory effect as enzyme action, and its concentration in culture medium was noticed to be closely related. In the presence of cuprous ion, activity of tyrosine is increased that causes an increase in betalain production. In different tissues of *A. hypochondriacus*, the activity of dioxygenases associated with Tyr resulted in the increment of betalain content. A further study is needed for temporal and spatial expression of tyrosine or dioxygenases and their regulation under biotic and abiotic factors. Both TYR and DOD activities as key enzymes have been detected to be coextensive with betalain content.

In *P. Americana* and *A. muscaria*, regulation of DOD is done at transcriptional level by means of basic helix-loop-helix (bHLH) and MYB that are responsive elements in the promoter region. MYB, bHLH, and WD40 repeats (WDRs) are three well-defined gene families known yet. These gene families encode transcription factors that interact with each other to produce MBW complex by which biosynthesis of flavonoids is regulated at the level of transcription. These transcription factors are also associated in the regulation of anthocyanin biogenesis. Anthocyanins and betalains have resemblance in function, biochemistry, and molecules which suggest a similar biosynthetic pathway of betalains as that of anthocyanins. In red beet, betalain biosynthesis is regulated by biosynthetic genes that utilize MYB-type transcription factor (BvMYB1) for regulation, which is in accordance with the above hypothesis. Betalain biosynthetic genes are activated selectively by BvMYB1. It causes interactions in MBW complex with anthocyanin bHLH transcription factors. It seems that BvMYB1 contains different residues than those that are responsible for interaction with bHLH transcription factors of anthocyanins. As a consequence, BvMYB1 will not bind to gene promoters responsible for anthocyanin biosynthesis that are also present in betalainic plants. Divergent nature of BvMYB1 shows that betalain biosynthetic genes are regulated by either independently or by MBW complex. Regulation by MBW complex will need an unknown bHLH transcription factor that is betalain specific with which BvMYB1 will interact for complex formation. However, it is to be discovered up till now how BvMYB1 interacts with genes to encode the increasing number of enzymes that catalyze the extended route of betalain production. Presence of BvMYB1 reinforces and suggests that anthocyanic and betalainic species share a common same ancestor and they control the production of selective pigments at the level of transcription. In addition to MYB and bHLH responsive elements, responsive elements of environmental stress are also present in the promoter region of DOD. In the production of betacyanins, regulation is provided by environmental stress which is revealed by studies of gene expression in *S. salsa calli*. Betacyanin production is associated with mRNA transcript levels of DOD, when treated with some growth regulators. Moreover, certain environmental factors such as UV light, sunlight, ratio of low red and far red light, salt, and oxidative and abiotic stress enhance the biosynthesis of betalains, while blue light suppresses the process as evident by biochemical studies. White light increases DOD expression, thereby increasing the production of betalains. As environmental factors have an influence on physiological factors, i.e., phytochrome, cytochrome P450, cryptochrome 2, Ca^{2+} signaling cascade, oxidants, transcriptional factors, etc., the biosynthesis of betalains is affected by environmental factors. How DOD promoter and physiological agents interact with each other is currently unknown. Different mechanisms of regulation by DOD can be studied easily because DOD is present in the form of a single transcript and there are numerous validated homologous sequences to examine.

Class II DNA transposable element is the latest physiological component associated with the regulation of betalains and is identified in *M. jalapa*. In the intron of CYP76AD3 that is a homologue of CYP76AD1, the sequence of DNA of the non-autonomous element dTmj1 was characterized. From the intron, deletion of dTmj1 causes red color diversification in yellow floral envelop (mutant). To discover more about the regulation of betalain biogenesis, gene sequences encoding enzymes of biosynthesis, expression during different developmental stages, and enzyme activity associated with certain ecophysiological factors, i.e., repressing agents and activating agents of enzymes in the pathway, should be searched more. Environmental stress raises B5GT also, which catalyzes the biosynthesis of betanins. It was first identified from *D. bellidiformis* and then further confirmed by reports on red beets. By oxidative stress, B5GT was induced in red beets, and an increase in betanin production was observed. Betacyanin biosynthesis was induced by reactive oxygen species (ROS) and that ROS was also found in *S. salsa* leaves after watering its seedlings with hydrogen peroxide and treatment of hydroponic cultures of red beet with copper sulfate.

By bacterial infection and wounding, B5GT can also be induced in addition to oxidative stress. These observations are coherent with UDP-glucosyltransferases of plant secondary metabolites' roles in upholding hormonal homeostasis, increasing water solubility and stability, compartmentalization, secondary metabolite regulation, and detoxification. B5GT participation in the biosynthesis of betalains stimulated by ecophysiological factors needs to explore more because it is clear now that in salt stress and dry conditions, the B5GT expression does not match with the biosynthesis of betacyanin in *A. hypochondriacus*. It seems that the expression pattern of B5GT is genotype and tissue specific [5].

4.14 Molecular Biology of Betalain Biosynthesis

Although betalains are of much more importance as natural food colorants, a small number of enzymes catalyzing the biosynthesis of betalains are purified and identified up till now. From higher plants, activities of only betanidin-glucosylating enzymes and betalain-degrading enzymes are described yet; a little research is made on synthetic genes. From a library of cDNA, which is formed by *Phytolacca americana*'s log-phase suspension culture, two clones were isolated that encode polyphenol oxidase. From the various organs of *Phytolacca* plants, temporal and spatial expressions were studied by Northern blot analysis of total RNA. Two clones have transcripts of 2.1 and 2.3 kb. In ripening of betalainic fruits, both transcripts were situated at significant levels. From *A. muscaria*, basidiomycete regulation and cloning of DODA gene were described. It was the first attempt to clone the betalain biosynthetic genes. From young specimens' cap tissues where biosynthesis of betalains had not been started, cDNA library was constructed. In *A. muscaria*, a single-copy gene encodes DODA as evident by Southern blot analysis. Mueller et al. transformed *Portulaca grandiflora* white petals with *A. muscaria* DODA gene that resulted in violet and yellow spots production that consisted of muscaflavin and betalain pigments. This pigmentation reveals that fungal enzyme DOPA dioxygenase (DODA) expresses itself actively in plants [2].

4.15 Biotechnology to Produce Betalains

Biotechnology explores cells, tissues, organs, or entire organisms by genetically handling them to synthesize desired compounds and developing them in vitro. This synthesis is advantageous than field cultivation because it does not depend upon environmental factors, geographical and seasonal changes, and political hindrance. Moreover, it permits self-imposed alteration of growth parameters, optimal and stable growth conditions, and constant quality control. Colorants developed by biotechnology are referred to as "natural colorants", due to which these are demanded mostly by customers. In in vitro culture, several plants such as *Portulaca grandiflora* Hook, *Myrtillocactus geometrizans* (Mart.) Console stem tissues, *Ph. americana*, *B. vulgaris* cell lines, tissues of *Chenopodium rubrum L.*, and others have been introduced for studying the biogenesis and betalain ultimate commercial production. In the table, the betalain yield of several synthesizing systems is given. For the production of pigments on commercial level, chemical synthesis of betalains is not practicable due to the involvement of several steps in the formation of betalamic acid and the low output of final products. By reacting the desired amine and betalamic acid at neutral or slightly acidic pH, betaxanthins can be produced in vitro non-enzymatically; in the acidic vacuole of plant, this happens immediately [4] (Table 4.1).

How does the regulation of TYR and CYP76AD1 and B5GT and B6GT occur? Is tyrosine, a starting material in the pathway, as a substrate not accepted by extracts of crude proteins obtained from yellow beet hairy root cultures [8]? Do betacyanins and betaxanthins arise from the same pool of betalamic acid? How does intracellular migration of betalains take place [5]. These are few questions related to betalain biosynthesis that are still unknown and to be discovered.

Few pathways are not known like major things such as biosynthetic routes and regulation also need further characterization, gene(s) involved in conversion of tyrosine to L-DOPA, and in the hydroxylation of tyrosine and performing biosynthetic regulation. Few pathways are known about the genes involved in pigment modification, but there is no any given authentic information about these pigments' transport into the vacuole and any possible pathway of degradation.

Enzyme involved for the conversion of tyrosine to L-DOPA is still unknown. Gene or protein identity is not reported for biochemical separation on betalainic. In vitro experiments and animal system examination have revealed that the formation

Table 4.1 Betalain pigment production by in vitro systems in plants [26]

Type of culture	Species	Pigment produced	Pigment content
Hairy root	*B. vulgaris* var. Boltardy	Betacyanins	0.7 mg/g fm
Hairy root	*B. vulgaris* var. Boltardy	**Betaxanthins**	1.3 mg/g fm
Cell suspension	*B. vulgaris* L. cv. Detroit Dark red	Betacyanins	22 mg/g dm
Hairy root	*B. vulgaris* L. var. Ruby Queen	Betacyanins	202mg/L
Callus culture yellow phenotype	*B. vulgaris* var. bikores monogerm	**Betaxanthins**	4.3μmol/g dm
Callus culture orange phenotype	*B. vulgaris* var. bikores monogerm	**Betaxanthins**	12.2μmol/g dm
Callus culture red phenotype	*B. vulgaris* var. bikores monogerm	Betacyanins	11.2μmol/g dm
Callus culture violet phenotype	*B. vulgaris* var. bikores monogerm	Betacyanins	28.0μmol/g dm
Transformed callus	*B. vulgaris* L. var. altissima	Betacyanins	13.8mg/g dm
Stem tissues	*Myrtillocactus geometrizans* (Mart.) Console	Betalains	---
Cell suspension	*Chenopodium rubrum* L.	Betcyanins	3–4mg/g dm
Cell suspension	*Portulaca grandiflora* Hook	Betaxanthins	---
Cell suspension	*Portulaca* sp.	Betacyanins	5.3 mg/g fm
Callus culture	*Mammillaria candida*	**Modified betaxanthins**	---

of L-DOPA from tyrosine is catalyzed by a PPO or tyrosinase-like enzyme. It was also supposed that the same enzyme is involved in the formation of L-DOPA and cyclo-DOPA. Discovery of CYP76AD1 (the beet R locus) and its requirement was made in the formation of cyclo-DOPA but not in the formation of L-DOPA (if it catalyzes its formation, then it is pleonastic with other genes). It is evident that only the R locus is required for catalysis of cyclo-DOPA formation reaction.

In modern days, cytochrome P450, or any other enzyme, including PPO-like enzymes, is involved for the conversion of tyrosine into L-DOPA. There are no reported mutants for tyrosine conversion to L-DOPA, which express that there is genetic dismissal for this activity.

In response to different stimuli, betalain synthesis is observed little, but to localize betalains in certain tissues, the activity of gene in this environmental or developmental regulation is unknown. By identification of regulatory genes, pathway evolution, environmental regulation, and developmental localization can be made clear. It will be advantageous for meliorating betalainic fruits and decorative plants by molecular or genetic ways. Expression of betalains responds to stresses in the same manner as anthocyanins. Its biosynthesis can be regulated by the anthocyanin regulatory network, by implying the co-option of regulatory genes. If it is possible, it can be an attracting process in which a regulatory network has shifted to regulate new genes and final products. In this case red pigment production and all the biological functions are still being performed by the new biosynthetic scheme.

Betalain pigments do not contain final compound. Many modifications occur from a small number of genes/enzymes to form a standard backbone. These may comprise of a shade-producing contributing factor and chromaticity of red/violet/pink or yellow/orange that occur in betalainic species. These modifications include only the glucose addition on cyclo-DOPA at two positions, either before or after it get condensed with betalamic acid.

Betalain synthesis in cell is unknown. The localization of membrane in the endoplasmic reticulum suggests due to the involvement of cytochrome P450 in the formation of cyclo-DOPA. Final products formed in vacuole are an enigma. Betalain pigments may have such transporters which are associated in anthocyanin pigment production and responsible for their movement in vacuoles. But their presence is unknown yet.

DODA-like genesis soluble and may exist in almost all life, but this protein type performs a new function in betalainic species. The genes of homologous biochemical functions in other plant species are unknown yet. Betalain pathway and non-pathway DODA sequence analysis has revealed two divergent themes at the catalytic site, one conserved in betalainic species and the other in non-betalainic species. It is suggested that DODA-specific biosynthetic ability depends upon the different motifs at the catalytic site. Conserved motifs may possibly be conserving due to phylogeny instead of its function, but this theory is not tested yet. It should be easily tested by using transgenic technologies and the current state of sequencing.

By introducing modern techniques of sequencing, analysis, and more genetic instruments and resources, the discovery of unknown genes and their functions can be made easier. Restricted presence of betalains in plants is a more interesting part of this pathway. The discovery of identified genes will give information on the function of betalain biosynthetic pathway and the way it was formed [12].

Some unanswered questions in tyramine-based biosynthetic pathway like obtainment of betalamic acid for condensation, mode of condensation of betalamic acid with tyramine or with dopamine (i.e., spontaneous or enzymatic), and spontaneous or enzymatic nature of intermolecular cyclization also arise. Enzymes are still unknown in the case of enzymatic described reaction [8].

4.16 Semi-Synthesis of Betalains

The semi-synthesis and purification of betaxanthins are based on the synthesis of Schiff base. Nitrogen atmosphere is used for carrying out the semi-synthesis of betalains which comprises of following steps:

4.16.1 Extraction of Starting Material

Betalain, dopaxanthin, betanin, and betanidin were extracted and purified from yellow and violet flowers of *Lampranthus productus* and also from red roots of *Beta vulgaris*. This purified betanin from *Beta vulgaris* (red beets) was used as a starting material for the semi-synthesis of betalains [27] and artificial betalainic coumarin, which was applied as a fluorescent probe for the live cell imaging of *Plasmodium*-infected erythrocytes [28].

4.16.2 Hydrolysis of Betanin

Betanin (0.2 mM) was subjected in basic medium, i.e., pH 11.4, to produce betalamic acid through hydrolysis.

4.16.3 Condensation

Betaxanthin was synthesized by the condensation of suitable amine betalamic acid produced by hydrolysis at pH 5.0 [27]. This newly produced substance reveals pseudo-betaxanthin, and its stability lasts for at least half year. The newly synthesized bond can be displaced by adding amines in aqueous solutions by using mild conditions over a wide range of pH at 25 °C. Under these conditions, betalamic acid is released, and related pigment is produced. This step successfully describes the process of semi-synthesis of betalains, i.e., betaxanthins and betacyanins [10].

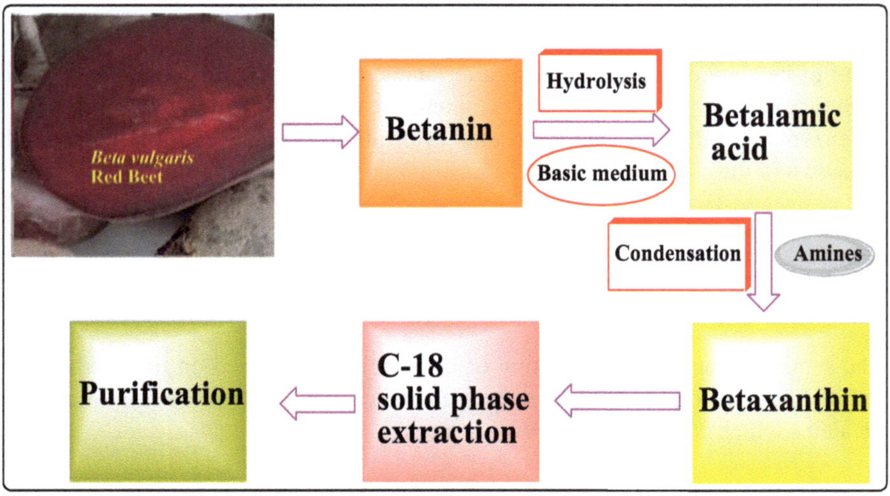

Fig. 4.9 Schematic representation of semi-synthesis of betalains

Most of the betalains were obtained as a result of immonium condensation of betalamic acid with the amines such as ethylamine, propylamine, N-methyl-ethanamine, 2-phenylethylamine, N-methyl-N-propylamine, aniline, pyrrolidine, N-methylaniline, N-ethyl-aniline, (S)-indoline-2-carboxylic acid, indoline, and (S)-phenylalanine.

4.16.4 Extraction and Purification

After the synthesis of betalains, extraction and purification of pigments were done.

4.16.5 C-18 Solid-Phase Extraction

C-18 solid-phase extraction was used for the extraction of synthesized pigments. In this extraction system, C-18 cartridges (1 mL) were conditioned with methanol (5 mL) and then by distilled water (10 mL) (Waters, Milford, MA, USA). Columns were rinsed with water to make samples free from buffers and salts. After removal of salts, conductivity of samples was lower than 1.0 mS/cm (pH/ C-900 conductivity detector, General Electric Healthcare, Milwaukee, USA).

By using ethanol, elution of betaxanthins was done and then concentrated under vacuum till dryness at room temperature. The residue was dissolved again in water for further use. The process gave 92–97% yield.

4.16.6 Automated Purification System

By using Akta purifier apparatus, betalains were subjected to anionic exchange chromatography (General Electric Healthcare, Milwaukee, USA). By using Unikorn software version 3.00 in a PC, the operation of apparatus was monitored. At three wavelengths, i.e., 280, 480, and 536 nm, elutions were monitored. Solvents were used as sodium phosphate buffer (10 mM) at pH 6.0 with NaCl (2 M). A 25 × 7 mm Fast Flow column with Q Sepharose (1 mL) purchased from General Electric Healthcare was used. It contained cross-linked agarose with quaternary ammonium to act as exchanger group, and its particle size was 90 μm. After the injection of sample, the elution was done with 0% solvent B (0.0–2.0 mL). The eluted band is washed, then a linear gradient was performed from 0% to 26% solvent B (over 15 mL), and 1 mL fractions are collected. 1 mL was the injection volume, and the flow rate was 1.0 mL per min [27] (Fig. 4.9).

References

1. Mohamed Yahya, K. (2006). *Studies on tissue culture system for the production of food colours from Beta Vulgaris L*, Doctoral dissertation, University of Mysore.
2. Delgado-Vargas, F., Jiménez, A. R., & Paredes-López, O. (2000). Natural pigments: Carotenoids, anthocyanins, and betalains—Characteristics, biosynthesis, processing, and stability. *Critical Reviews in Food Science and Nutrition, 40*(3), 173–289.
3. Pavokovic, D., & Krsnik-Rasol, M. (2011). Complex biochemistry and biotechnological production of betalains. *Food Technology and Biotechnology, 49*(2), 145.
4. Davies, K. (Ed.). (2009). *Annual plant reviews, plant pigments and their manipulation* (Vol. 14). Oxford/Boca Raton: Blackwell Publishing/CRC Press, Boca Raton.
5. Khan, M. I., & Giridhar, P. (2015). Plant betalains: Chemistry and biochemistry. *Phytochemistry, 117*, 267–295.
6. Gandía-Herrero, F., & García-Carmona, F. (2013). Biosynthesis of betalains: Yellow and violet plant pigments. *Trends in Plant Science, 18*(6), 334–343.
7. Delgado-Vargas, F., & Paredes-López, O. (2002). *Natural colorants for food and nutraceutical uses*. Boca Raton: CRC Press.
8. Xiao-Hong, H., Zhao-Jian, G., & Xing-Guo, X. (2009). Enzymes and genes involved in the betalain biosynthesis in higher plants. *African Journal of Biotechnology, 8*(24), 6735.
9. Sakuta, M. (2014). Diversity in plant red pigments: Anthocyanins and betacyanins. *Plant Biotechnology Reports, 8*(1), 37–48.
10. Esatbeyoglu, T., Wagner, A. E., Schini-Kerth, V. B., & Rimbach, G. (2015). Betanin—A food colorant with biological activity. *Molecular Nutrition & Food Research, 59*(1), 36–47.
11. Mueller, L. A., Hinz, U., & Zryd, J. P. (1997). The formation of betalamic acid and muscaflavin by recombinant DOPA-dioxygenase from amanita. *Phytochemistry, 44*(4), 567–569.
12. Chen, C. (Ed.). (2015). *Pigments in fruits and vegetables: Genomics and dietetics* (pp. 127–140). New York: Springer.
13. Moreno, D. A., García-Viguera, C., Gil, J. I., & Gil-Izquierdo, A. (2008). Betalains in the era of global Agri-food science, technology and nutritional health. *Phytochemistry Reviews, 7*(2), 261–280.
14. Stintzing, F. C., & Carle, R. (2004). Functional properties of anthocyanins and betalains in plants, food, and in human nutrition. *Trends in Food Science & Technology, 15*(1), 19–38.
15. Gengatharan, A., Dykes, G. A., & Choo, W. S. (2015). Betalains: Natural plant pigments with potential application in functional foods. *LWT-Food Science and Technology, 64*(2), 645–649.
16. Gandía-Herrero, F., Escribano, J., & García-Carmona, F. (2005). Betaxanthins as substrates for tyrosinase. An approach to the role of tyrosinase in the biosynthetic pathway of betalains. *Plant Physiology, 138*(1), 421–432.
17. Tanaka, Y., Sasaki, N., & Ohmiya, A. (2008). Biosynthesis of plant pigments: Anthocyanins, betalains and carotenoids. *The Plant Journal, 54*(4), 733–749.
18. Herbach, K. M., Stintzing, F. C., & Carle, R. (2006). Betalain stability and degradation—Structural and chromatic aspects. *Journal of Food Science*, (4), 71, R41.
19. Jain, G., & Gould, K. S. (2015). Are betalain pigments the functional homologues of anthocyanins in plants. *Environmental and Experimental Botany, 119*, 48–53.
20. Harris, N. N., Javellana, J., Davies, K. M., Lewis, D. H., Jameson, P. E., Deroles, S. C., et al. (2012). Betalain production is possible in anthocyanin-producing plant species given the presence of DOPA-dioxygenase and L-DOPA. *BMC Plant Biology, 12*(1), 34.
21. Khan, M. I. (2016). Plant betalains: Safety, antioxidant activity, clinical efficacy, and bioavailability. *Comprehensive Reviews in Food Science and Food Safety, 15*(2), 316–330.
22. Strack, D., Vogt, T., & Schliemann, W. (2003). Recent advances in betalain research. *Phytochemistry, 62*(3), 247–269.
23. Sánchez-Ferrer, Á., Rodríguez-López, J. N., García-Cánovas, F., & García-Carmona, F. (1995). Tyrosinase: A comprehensive review of its mechanism. *Biochimica et Biophysica Acta (BBA)-Protein Structure and Molecular Enzymology, 1247*(1), 1–11.

24. Lipscomb, J. D. (2008). Mechanism of extradiol aromatic ring-cleaving dioxygenases. *Current Opinion in Structural Biology, 18*(6), 644–649.
25. Wang, C. Q., Song, H., Gong, X. Z., Hu, Q. G., Liu, F., & Wang, B. S. (2007). Correlation of tyrosinase activity and betacyanin biosynthesis induced by dark in C 3 halophyte Suaeda salsa seedlings. *Plant Science, 173*(5), 487–494.
26. Georgiev, V., Ilieva, M., Bley, T., & Pavlov, A. (2008). Betalain production in plant in vitro systems. *Acta Physiologiae Plantarum, 30*(5), 581–593.
27. Gandía-Herrero, F., Escribano, J., & García-Carmona, F. (2010). Structural implications on color, fluorescence, and antiradical activity in betalains. *Planta, 232*(2), 449–460.
28. Stafford, H. A. (1994). Anthocyanins and betalains: Evolution of the mutually exclusive pathways. *Plant Science, 101*(2), 91–98.

Chapter 5
Role of Betalain in Human Health

Various bioavailable compounds from plants have recently attracted much attention as they are vital protecting factors in human health. Pharmacological activities of betalain such as antioxidant, anticancer, anti-inflammatory, antilipidemic, and antimicrobial make them able to play a central role in human health. When betalains become part of regular diet in humans, they offer fortification against stress-related syndromes due to their ability to obstruct lipid peroxidation and oxidation other than its color-imparting features. Several pathological syndromes are well treated and cured by health-promoting agents like beetroot. Beetroot supplementation containing betalain is recounted to attenuate inflammation, decrease blood pressure, avoid oxidative burden, restore cerebrovascular hemodynamics, and block endothelial function in humans [1].

5.1 Oxidation of Low-Density Lipoproteins

The antioxidant activity of phytochemicals containing polyphenol compounds is a noteworthy issue that explains the beneficial effect of food on humans. Betalain is one of the compounds that are obtained from fruits and vegetables of different plants. Its enriched low-density lipoprotein (LDL) fraction was obtained by spearing of plasma with both purified betanin and indicaxanthin, which then led to the separation of LDL in humans. Maximum binding of both compounds is about 0.50 mol/mg LDL protein. Activated phagocytes, secreted by hemeprotein and myeloperoxidase (MPO), are part of the protection process against attacking nitrite and pathogens that is one of the mode of LDL oxidation. Enormous oxidation of these lipids has been proposed by nitrogen peroxide, which involves an electron oxidation of nitrite through peroxide-activated myeloperoxidase. It was observed that betanin, at micromolar concentrations, prevents the fabrication of lipid hydroperoxides in living organism where low-density lipoprotein acquiesced to a MPO or nitrite-promoted oxidation (Fig. 5.1). Marvelously, it stood prominent that yet

E. Akbar Hussain et al., *Betalains: Biomolecular Aspects*,
https://doi.org/10.1007/978-3-319-95624-4_5

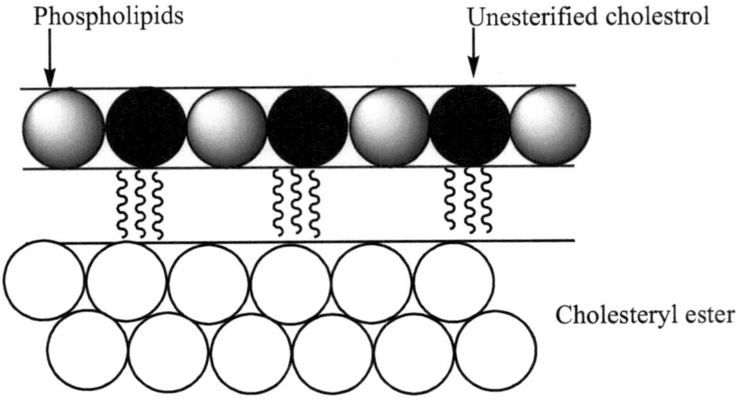

Fig. 5.1 Low-density lipoprotein

unrevealed, oxidation products of betanin through nitrite or MPO can constrain either species that induced LDL oxidation as successfully as the original parent compound.

Gene regulatory activity is partly displayed by betalain via Nrf2-dependent signaling routes. The scavenging activity to reactive oxygen species of betalain was prompted by phase II enzymes and antioxidant defense mechanisms. LDL oxidation and DNA damage are also prevented by betalains. Potential blood pressure-lowering effects of betalain are chiefly mediated by dietary nitrates. Strong health-promoting potential has been exhibited by dietary betalains, as they inhibit the formation of tumors in vivo in mice.

Betalains guard erythrocytes against hemolysis that prompts the autogenous glutathione synthesis in human erythrocytes. LDL particles are shielded by betalains against oxidation due to their cationic structure and intermingle with the polar ends of the LDL constituent. When betalain was compared to other antioxidants like catechin and α-tocopherol, they countered LDL oxidation at relatively low concentrations. In humans, they reduced the biomarkers of lipid oxidation such as F2-isoprostanes (in plasma), malondialdehyde (in plasma), and lipid hydroperoxide (in LDL). Reduction of DNA damage tempted by hydrogen peroxide in human lymphocytes is mediated by *Opuntia ficus-indica*, which is derived from betalains. Prospective well-being benefits of betalain-rich foods are represented in Fig. 5.2.

5.2 Effects of Betalains on Different Body Organs

5.2.1 Shielding Effect of Betalain on the Heart

Betalains present in beetroots perform shielding function for the heart. It lowers the homocysteine (Fig. 5.3) level. High homocysteine levels are related to blood vessel inflammation. They also lower the C-reactive protein (CRP) that damages the arteries.

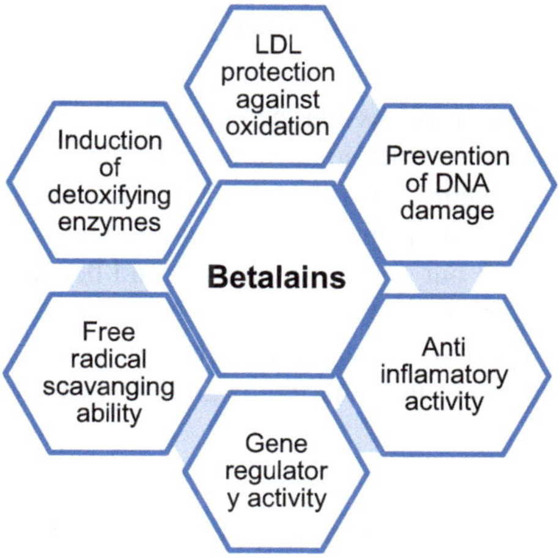

Fig. 5.2 Potential health benefits of betalains

5.2.2 Loss of Fatty Deposits from the Liver

Fatty deposits in the liver are reduced by taking betalains in food. These deposits are due to diabetes, alcohol abuse, and conditions like protein deficiency and obesity. Liver cell function is stimulated by increasing the production of detoxifying enzymes in the liver (catalase and glutathione). When betalains enter the human body, they restore strength on a cellular level by reducing any toxins adjoining the cells that enable useful nutrients to reach the cells. Inflammations in our body can be reduced by using betalains. Intake of beetroot juice is beneficial during jaundice because it provides the desired fluids and carbs in this situation.

5.2.3 Effect on Digestion

Beetroot as a natural laxative helps in treating constipation and hemorrhoids. Digestion rate is multiplied by taking beetroot juice so that the secretion of digestive acids becomes high.

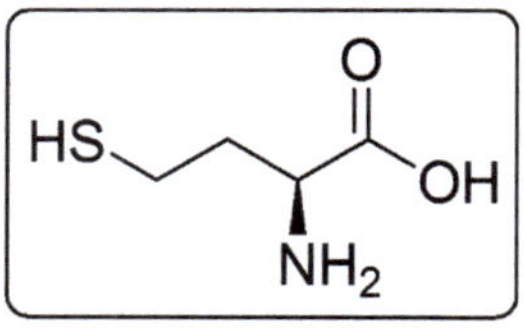

Fig. 5.3 Homocysteine

5.2.4 Action as a Guard to the Skin

Skin problems like acne are cured by betalain due to its astonishing detoxifying abilities. Nutritional value of betalain is responsible to refresh and nourish the skin by keeping it hydrated. Premature aging is also controlled by its excessive use. Many of the impurities, toxins, radiation, and other injurious influences on our body affect our skin. Antioxidants and betalains are required to shore up the defensive layer. Gaps of our cells can be filled (including the cells in our skin) and converted to the strong, vigorous nature of young cells at any age. They keep the skin protected from withering by giving a healthy atmosphere to the cell [2].

5.2.5 Enhancing Nerve Effect on Eye Health

Betalains found in Swiss chard help in nerve signaling that is responsible for the communication between the eyes and brain by protecting the health of the nervous system.

5.2.6 Human Red Blood Cells

Transition metals endorse oxygen free radicals due to increased oxygen tension and high concentration of iron. Additionally, the oxidative modifications of the RBC membrane follow the injury to the endothelium cells. It was stated that ex vivo upsetting of blood in strong and fit human beings with pure indicaxanthin or betanin gave a full incorporation of betalains in red blood cells to equivalent extent (approximately 1 nmol/mL of packed cell). Oxidative strain effects in red blood cells are usually reduced by bioavailable phytochemicals. The comparison of cumene hydroperoxide and betalain resulted in more resistant erythrocytes upon release and prompted oxidative hemolysis than those homologous that are not enriched erythrocytes.

5.2.7 Strong Skeletal System by Betalain-Rich Food

Inflammation-relieving natural foods such as betalain are also used for arthritis relief. Prickly pear is the solitary plant to comprise 24 recognized betalains that are persuasive anti-inflammatory agents. They give prickly pears their purple, red, and yellow color. Calcium in prickly pear is responsible for bone growth. In fact, 99% of the calcium in your body is found in your bones and teeth. Polyphenolic pigments found in beetroot are betalains. Swiss chard (impressive vegetables) is also one of the best sources of betalains. Two key nutrients vitamin K and calcium in Swiss chard maintain a strong skeletal structure and mineral density to avert weak bones from fractures [3].

The best organic foods as juice are beetroots which are used in raw juice therapy that can alleviate symptoms of arthritis and other bone and joint disorders such as gout. Beetroot juice also has beneficial effects on bone diseases such as brittle bone disease and osteoporosis. Bones are made up of salt of calcium and phosphorus as mineral constituent, and beetroot juice is accompanied with mineral calcium.

5.2.8 Effect on Human Endothelial Cells

Vascular endothelial cells are a direct mark of pro-inflammatory stimuli, suffering from numerous redox-induced signaling pathways, leading to the manufacture of chemotactic aspects, lipid mediators, and cytokine adhesion compounds like adhesion molecule-1, which are communicated by vascular endothelial cells in the category of inflammatory situations. Activity of indicaxanthin and betanin in controlling the appearance of cytokine adhesion molecules is tested through in vitro model of inflammation comprising of umbilical vein endothelial cells (HUVEC) motivated by the pro-inflammatory cytokine tumor necrosis factor-α (TNF-α). ICAM-1 expression at micromolar concentrations is inhibited by both pigments. In bringing together the antioxidant characteristics of the molecules, there is a potential pharmacological attention in pathologies, for instance, atherosclerosis, a complex procedure to a great degree intervened by cytokines, growth factors, adhesion molecules, and composites involved in redox-sensitive monitoring mechanisms, or other inflammatory ailments categorized by tissue degeneration because of endothelial dysfunction such as atherothrombosis, lower-limb ischemia, and stroke [4].

5.2.9 On Microsomal Membranes

The attraction of betanidin or betanin for microsomal membranes was exposed through the estimation of the degree of movement of molecules by dialysis tube, though in the occurrence or absence of microsomes. The antiradical action of various betanin in this atmosphere has been judged by accepting their absorption to membranes as ferrous chloride or ascorbate or hydrogen peroxide stimulated by myoglobin. Though attributable to its electron-giving commotion, low amount of betanin was prooxidant in the reactions catalyzed by iron/ascorbate due to the discount of ferric into the prooxidant ferrous ion. An actual reserve of lipid peroxidation necessarily increased the concentrations up to good extent.

5.2.10 Cure of Pregnancy-Related Problems

High folate contents in beetroot juice are beneficial for expectant mothers which in turn help in preventing birth defects. Iron-rich beetroots help in avoiding anemia during pregnancy by increasing hemoglobin and blood volume. Cattle nurtured with huge amounts of sugar beet leaves exhibit infertility and irregularities of the genital tract [1].

5.2.11 Effect on the Immune System of Mice

Betalains from red beets were radioprotective in mice. Irradiation with γ-rays brought a marked decrease in white blood cell amounts and spleen and thymus index and a clear increase in the micronucleus rates of polychromatic erythrocytes in the bone marrow of these mice. Betalains partially reinstated these values in a dose-subjected mode. γ-Ray radiation also initiated lipid oxidation in the liver, spleen, and kidney as imitated by high levels of MDA.

Piper longum (Pippali) fruits were used in the form of ethanolic extract to defend mice against the radiation-induced decline in white blood cells, bone marrow cells, and α-esterase-positive cells. Bone marrow tissues are very sensitive to radiation and cause chromosomal abnormalities and increase the micronucleus rates of polychromatic erythrocytes. The irradiation-induced variations in micronuclei and chromosomal anomalies are dose dependent. The administration of betalains from red beets in literature incompletely prohibited the growth in the micronucleus rate of polychromatic erythrocytes in the bone marrow of mice, indicating that betalains from red beets are antimutagenic.

5.3 Effect of Betalain Against Different Diseases

5.3.1 Lowering of Blood Pressure

A study had been published in the *Journal of the American Heart Association* that blood pressure of a person instantly lowered within 24 h after consumption of beetroot juice (250 mL) and nitrate tablets. Nitrates found in beetroots transformed into nitrites by the salivary action in order to lower blood pressure [5].

5.3.2 Reduction of the Blood Cholesterol Level

Nitrate compounds found in beets depress the cholesterol level. Beetroot fibers along with its betalain containing folate lower blood triglyceride and blood cholesterol level.

5.3.3 Deleterious Effect on Cancer Cells

Betalains are disease-fighting phytonutrients, and their extract from beet functions as cancer-protective antioxidants and anti-inflammatory bits. Cancers such as lung, stomach, colon, and breast are locked by its use. Growth of CD8 cells is recognized by betalains which reduce the rate of proliferation of tumor cells in the colon, breast nerves, testicle, stomach, and lungs. It also has the ability to detect and remove abnormal cells [6].

5.3.4 Toxicity Prevention

Direct effects of betalain on the liver provide noteworthy protection of the body from toxins in air, water, and food. Toxin fixation serves to combat these venoms [7]. Activity of CYP-2E1(a special enzyme that neutralizes daily toxins) is increased; betalains protect the vital organs from carcinogens, hormone disruptors, respiratory toxins, developmental toxins, and neurotoxins that are responsible for one quarter of all the world's diseases.

Medical specialists all over the world are identifying the risk of health modeled by inflammation. Increasing inflammation, which occur as a result of toxins stabbing system and also when infection sets in subsequent to an injury, is significant in our effort on the way to further difficulties. An extensive variation of ailments from stroke to arthritis and heart disease is seen if inflammation is not checked. By taking betalains, inflammation in the body can reduce, and other anti-inflammatory markings work directly on the fluid neighboring cells. There, they tap off the toxins and extra water to lessen inflammation and rebalance the cellular environment.

5.3.5 Anemia

Thirty-seven percent of the daily requirement of folate (also rich in iron 6% of daily requirement) can be provided by betalain in beetroot juice to overcome anemia. The combination of iron (14% RDA) with the antioxidants (127% RDA) in beet makes it a cherished source of iron and also a strong oxidant that permits oxygen to enter in the blood. Beet juice aids in anemia reversal. To treat anemia, enough vitamin B12 is found in animal food and some sea algae (e.g., E3 Live and marine phytoplankton). It also helps in building up the hemoglobin.

5.3.6 *Thalassemia*

β-Thalassemia is a genetic hemolytic sickness first activated by hemoglobin auto-oxidation and precipitation and then characterized by high generation of interactive oxygen species. Cell components are damaged due to depletion of the RBC antioxidant defense, injury of morphology and property of cell membrane, and enhanced RBC destruction. β-Thalassemia is treated by antioxidant vitamins and phytochemicals. Defensive possessions of indicaxanthin on both membranes and solvable sections of β-thalassemic RBCs accede to an in vitro oxidation by cumene hydroperoxide that is established. Lipid hemolysis is prevented by betalain (dose dependently) by retarding vitamin E and GSH depletion. It was also described that upsetting of blood in thalassemia patients with indicaxanthin brings about its incorporation in RBCs, representing pathological changes to the membrane that does not affect the trans-bilayer drive of this phytochemical. Therapeutic interests of this phytochemical are because of incorporated indicaxanthin in the redox equipment of β-thalassemia RBCs [8].

5.3.7 *Effect on Calcium-Related Diseases*

Calcium absorption ability of beetroot juice makes it able to treat conditions like osteoporosis. Inorganic calcium can prevent problems like kidney stones, gout, arthritis, and atherosclerosis as it is not deposited in the blood vessels.

5.3.8 *Detoxification of the Blood*

Betalains because of glutathione presence are able to detox and cleanse the blood from toxins, heavy metals, and waste that makes them water-soluble and exerted through urine. It is flushed out from body; that's why it is known as a natural "blood cleanser." Fibers in beets help to "sweep" the digestive tract of waste and toxins for regular bowel movements. The blood of animals becomes purified and free from toxins and mutagens when beet extract is injected in them.

5.3.9 *Reduction in Tumor Cell Lines*

Betanin, at different quantities ranging from 12.5 μg/mL to 200 μg/mL, indicated growth-dependent retardation against colon, breast, stomach, and lung tumor cells and central nervous system at concentrations varying 12.5–200 μg/mL. The combination of betanin with anthocyanin resulted in a remarkable reduction in cell growth.

Detoxification of electrophilic carcinogens in cells is carried out through phase II enzymes, such as quinone reductase during the early phases of a cancerous procedure. Studies on murine hepatoma cells exposed betanin as quinone reductase inducer.

5.3.10 Prevention of Birth Defects

100 grams of beetroot provides 27% of the RDA for folic acid to stop numerous birth problems in babies.

5.3.11 Beeturia (Red Urine)

Red urine was observed after ingesting of beetroot. In any form, beeturia is not harmful as it is just an indication of a problem in the iron metabolism in the body. A health checkup in relation to iron metabolism is suggested.

5.4 Relations of Betalains with Hemeproteins

Basically hemeproteins are oxygen carriers. Redox grade of the heme iron permits these molecules to achieve their function. Hemeproteins are represented as applicants for persuading oxidative stress due to their peroxidase activity or similar potential in the presence of hydrogen peroxide or normal alkyl hydroperoxides. Extremely reactive reaction intermediates with hypervalent iron are produced by reacting hemeproteins with oxygen contributors which can be reestablished by two consecutive electron-shifting reactions to the heme edge from suitable electron-enrich substrates. The interface of hemeproteins with betalains was first indicated in literature that considered oxidation products resulting from the reaction of betanidin plus betanin with sterile horseradish peroxidase which display their health-stimulating activity. Heme iron may be oxidized under powerful oxidant pathological situations; this improved form is perferryl-Hb, which was an intermediate in the oxidative degradation of hemoglobin. The spectrophotometric analysis is carried out that on the molar basis, the reducing action of indicaxanthin toward the hypervalent heme iron of perferryl-Hb is one order of size upper than the demonstrated by well-recognized reductants like trolox and ascorbate. Dietary phytochemicals can also act in the gastrointestinal tract which is linked to the heme proteins, scavenging highly oxidizing hypervalent-iron myoglobin (perferryl-Mb) produced during meat ingestion that may preserve oxidizable lipids and evade the production of potentially toxic lipid hydroperoxides.

Myeloperoxidase (MPO), one of the human heme enzymes, shows crucial parts in the protection against attacking pathogens through oxygen-reliant antimicrobial property by converse. MPO has huge potential to execute harm to host tissues by its capability to speed up the formation of a complex array of reactive oxidants, nitrogen dioxide, counting organic free radicals, drug betanin and hypochlorous acid, and indicaxanthin. At very slight micromolar concentrations, they were exposed to interfere in the catalytic cycle of MPO by the reduction of hypervalent heme iron produced on the oxidation of ferric MPO via hydrogen peroxide. Moreover, betalains were able to forage HOCl mediators of inflammation catalyzed by cyclooxygenase enzymes, COX-1 and COX-2; micromolar concentrations of betanin were originated to inhibit the activities of COX-1 (33%) and COX-2 (97%).

5.5 Biomimetic Membrane

Literature relevant to biological membranes under controlled conditions can be carried out by liposomal oxidation model of 1-stearoyl-2-linoleoyl-sn-glycerol-3-phosphocholine, using fluorescence spectroscopy permitted to accomplish physical and chemical influences of reactivity and interfaces of different compounds with two layers of lipids. Lipid oxidation is prohibited by indicaxanthin when combined with liposomal bilayers of phosphatidylcholine which endured oxidation through water-soluble azo compound. Kinetic study suggested that the position of indicaxanthin in the bilayer may allow shooting of radicals from the aqueous stage, along with lipoperoxyl radicals prepared in the lipid. Antioxidant potential of the phytochemical such as regeneration of indicaxanthin is enhanced from its radical. Antioxidant activity of betanin was measured by using a liposomal oxidation model of 1-stearoyl-2-linoleoyl-sn-glycerol-3-phosphocholine through fluorescence spectroscopy. This oxidation was introduced by ferrous chloride linked to recognized marketable antioxidants (BHA, BHT, TBHQ, all at a 10 μM level), used as standard, betanin at 180 μM level displayed 71% inhibition.

References

1. Wootton-Beard, P. C., & Ryan, L. (2011). A beetroot juice shot is a significant and convenient source of bioaccessible antioxidants. *Journal of Functional Foods, 3*(4), 329–334.
2. Strack, D., Vogt, T., & Schliemann, W. (2003). Recent advances in betalain research. *Phytochemistry, 32*, 47–69.
3. Sapers, G. M., & Hornstein, J. S. (1979). Varietal differences in colorant properties and stability of red beet pigments. *Journal of Food Science, 44*(4), 1245–1248.
4. Marchant, D. J., Boyd, J. H., Lin, D. C., Granville, D. J., Garmaroudi, F. S., & McManus, B. M. (2012). Inflammation in myocardial diseases. *Circulation Research, 110*(1), 126–144.

5. Omar, S. A., Webb, A. J., Lundberg, J. O., & Weitzberg, E. (2016). Therapeutic effects of inorganic nitrate and nitrite in cardiovascular and metabolic diseases. *Journal of Internal Medicine, 279*(4), 315–336.
6. Kapadia, G. J., Rao, G. S., Iida, A., Suzuki, N., & Tokuda, H. (2011). *Synergistic cytotoxic effect of red beetroot (Beta vulgaris L.) food colorant E162 with anticancer drug doxorubicin (Adriamycin) against human prostate, breast and pancreatic tumor cells*. American Society of Pharmacognosy, 241.
7. Cai, Y., Sun, M., & Corke, H. (2003). Antioxidant activity of betalains from plants of the Amaranthaceae. *Journal of Agricultural and Food Chemistry, 51*(8), 2288–2294.
8. Raven, P. H., Ray, F., & Susan, E. (2004). *Eichhorn biology of plants* (p. 465). New York: W. H. Freeman and Company.

Chapter 6
Bioactivities of Betalains

Betalains, the group of chemicals responsible for color in red beet, are found to exhibit powerful antioxidant activity and free radical scavenging properties with potential health benefits in humans, such as boosting immune system and prevention of cardiovascular diseases, neurodegenerative disorders, and cancer [1].

6.1 Free Radical Scavenging Activity

Antiradical activity is generally a measure of the ability to hang up or inhibit oxidation of carbohydrates, lipids, proteins, and DNA which are generated by RNS and ROS that are reactive nitrogen species and reactive oxygen species, respectively. Due to their biological activity and RSA, antioxidants are referred as insurance to health. Delay or prevention of proteins, lipids, carbohydrates, and DNA oxidation induced by free radical species of oxygen and nitrogen is known as antiradical or antioxidant activity. Antioxidants ensure good health due to their anticipated free radical hunting activity and other biological activities. There is growing awareness in betalains since the last 15 years as their antioxidant activity was observed initially in 1998. Studies revealed the relationship between antioxidant activity, crude plant extracts, and betalain contents projected the possibility of using these extracts as biologically active food essences. Studies revealed that at neutral or basic pH, the efficiency of betanin to scavenge radicals was doubled than anthocyanins. Betacyanin fraction showed enhanced trapping activity for free radicals than flavonoid fractions obtained from red pitaya [2].

RSA investigation involves a broad category of betalain concentrations at various pH conditions. Moreover betaxanthin antioxidant power against DPPH at pH 7 was higher than betacyanins. That is why there is no suitable or adequate RSA that truly signifies the actual antioxidant ability of betalains. Its stability is strengthened by acylation and glucosylation meaning that neither miraxanthin V nor betanidin will be stable enough to exhibit their RSA in product form. Furthermore, the RSA

E. Akbar Hussain et al., *Betalains: Biomolecular Aspects*,
https://doi.org/10.1007/978-3-319-95624-4_6

of many betalains, excluding betanidin, were shown to increase with pH, but on the other hand, stability declined in the range 3 > pH > 7. Hence, while characterizing the RSA, chief consideration must be given to factors responsible for stability in order to present a comprehensive picture.

Preceding evidences that resulted from an in vitro study showed that betanidin and betanin are proficient retarders of LDL oxidation and membrane lipid peroxidation. Betanin's proficiency was higher than betanidin. Moreover, betanin revealed moderate to excellent inhibition of lipid peroxidation in a liposome system. This reflects that betanin and indicaxanthin are resistant to the oxidative damage of LDL.

Indicaxanthin, owing to its noticeable bioavailability, was highly effective. It was observed that in the alteration in the cellular environment after removal of betalains, these mechanisms become more supported. Due to its RSA, betalains quench free radicals, while cellular environment settled the oxidative pressure inserted by toxicants so that the redox balance is restored. Other than these possible functions, betanin as well as indicaxanthin may act as a signal in vivo to regulate and to carry out epigenetic modulation of gene expression, respectively, to suppress tumor generation via redox-sensitive transcription like NF-κB, activator. It also prevents the formation of AGEs by inhibiting the Maillard reaction. There is no proper information about the in vivo role of transformation products of betalain in literature so far.

With the increase in hydroxyl groups and hydrogen donors (i.e., imino groups), betalain radical scavenging properties tend to increase. In radical scavenging properties, the presence of catechol in betalains seemed to be very important. Moderate radical scavenging activity is shown by betaxanthins due to the absence of phenolic hydroxyl groups. Radical scavenging activity is reduced by glycosylation of betalains. 6-*O*-glycosylated betacyanins showed more radical scavenging action than 5-*O*-glycosylated betacyanins; this fact shows the influence of glycosylation on radical scavenging potential [3]. Betalains with cyclic amine group show similarity with ethoxyquinone and have relevant radical scavenging properties.

Betalains are an excellent source of antioxidants as revealed by several studies [4]. Two molecules of Fe^{3+} are reduced to Fe^{2+} by each molecule of betalamic acid. We can say that betalamic acid donates two electrons to an oxidant in order to reduce it. An ancient study revealed that an increase in blood pressure and heart rate is caused by betanin injection; thus it affects cardiovascular system. These effects were not observed by oral cannulation of betalains because in this case gastrointestinal tract is degraded. Due to these observations, scientists were interested in antioxidant activity that is related to physiological activities in the body. The protection against lipid peroxidation is an important factor to ensure a substance as an antioxidant in vivo [2].

Several studies revealed that betalain dyes resist oxidative injury of cellular components [5]. Two metabolites of betalains, betanin and betanidin, are responsible for decreasing the linoleate damage and lipid membrane oxidation persuaded by cytochrome C oxidase, hydrogen peroxide, oxidized myoglobin, and free ions. Its high radical neutralizing activity seems due to its special power to donate electrons and ability to neutralize radical leveling most reactive cell membranes [5, 6]. Betacyanin and betaxanthin antioxidant capacity is measured by modified method using 1,1-diphenyl-2-picrylhydrazyl, and results revealed that both of them have four times greater antioxidant activity than ascorbic acid, polyphenols, and catechin [7]. Antioxidant compounds as well as a complex mixture of natural materials are present in vegetable juices that enhance antioxidant capacity and are thus good for health. In resisting cardiovascular diseases, neurological disorders, and cancer, bioactive compounds in fruits and vegetables play an important part. Beetroot is among the top ten antioxidant-containing vegetables. Imino group, hydroxyl group, and phenolic moieties are responsible for antioxidant activity of beetroot juice. Betalains in beetroot stop induction of oxygen radical and oxidation of molecules in biological system [8]. Betalain extracts obtained from mature beetroot have lower antioxidant activity than immature root extract. Due to the high concentration of phenolic compounds, hairy root extracts have high antioxidant activity and collaborative effects with betalains. In health issues like atherosclerosis, obesity, and cancer which are due to oxidative stress, betalains could be beneficial. Oxidative metabolism of neutrophile is responsible for the initiation of characteristic immunity and is inhibited in obese individuals. The in vitro experiments had shown proapoptotic effects on stimulated neutrophils by use of *B. vulgaris* root juice. By using differential spectrophotometry, antioxidant capacity of betalains was measured by following the method developed by Nilsson in 1970 [7]. By using phosphate buffer of pH 6.5, extracted juices were diluted and then mixed with [9] diammonium salt [10] of [2,2′-azinobis-(3-ethylbenzthiazoline)-6-sulfonate] (ABTS) solution of cation radical [9, 10], and then its absorbance was measured at 476 nm for betaxanthin (yellow) and at 538 nm for betacyanin (red) pigments after 6 min. Antioxidant activity (Mmol Trolox equivalent/mL) was measured with the help of Helios Alpha spectrophotometer (Thermo Electron Corporation, USA) with a temperature-controlled cell holder of Thermo Electron Corporation of USA. Statistical analysis was done by using STATISTICA ver. 7 software packages. Mean values, standard deviations, and correlation coefficients were determined statistically. By applying least squares method, regression was calculated between TEAC values of beetroot juice and pigment contents. There was a high variation between pigment contents and antioxidant capacity in cv. Czerwona Kula beets' juices from 2002 cultured crops. High variation in contents of pigments and antioxidant capacity was observed in juices from the harvest of 2002. Red pigment contents were ranged from 0.57 to 1.37, while yellow pigments were in the range of 0.40–0.61 mg/mL. The antioxidant capacity is directed in the range of 10.6–21.7 μmol Trolox/mL [9].

By ingesting only a specific amount of red beet extract, there was intensification in antioxidant level observed in human body. In irradiated mice, oral intake of red beet betalain contents (5, 20, or 80 mg/kg body weight) tends to enhance the capacity of glutathione peroxidase and superoxide dismutase (antioxidant enzymes) in the organs involved in metabolism. This increase in activity depends on the amount of dose intake. Betalains can change the key of ultimate imbalance between species resulting oxidation and organism's antioxidant defense system are able to produce a suitable environment in cells to oppose oxidative stress [11].

Radical scavenging activity of betalain samples had been measured by modified DPPH method. The standard solution (1 μM) was prepared by taking 22 mg of DPPH and dissolving it in 50 mL of methyl alcohol. It was stored at −20 °C, 6 × 10–5 M. Working solutions were made by adding 6 mL standard solution in 100 mL of methanol in order to get 0.8 ± 0.02 value of absorbance at 515 nm by using a spectrophotometer. For 30 s, 0.1 mL extracts of both were twisted with 3.9 mL DPPH standard solution and permit to react for 30 min. After this duration, the absorbance was recorded at 515 nm. A control was also analyzed without adding extract. The DPPH solution without extract was evaluated as control, and trapping activity was calculated by standard formula [12].

The oxidation of compound linoleate through cytochrome C was jammed by betanin, catechin, betanidin, and α-tocopherol where IC_{50} values are in the range of 0.4–5 μM. In another study, the IC_{50} values for soya bean lipoxygenase via betanin, betanidin, and catechin were recorded to be 0.6, 0.3, and 1.2 μM, respectively. This investigation relieved that betalains might be enough strong antioxidants as compared to catechins and some other flavonoids. Due to natural antioxidants, they may provide shield against oxidative strain-related disorders. That is why consumers will be aided from regular usage of products that are rich in betalains like red beet juice and its related products or food colors having betalains as nontoxic natural colorants [13].

6.2 Anti-Inflammatory Effects

Inflammation involves recruitment of plasma and leukocytes (components of blood) at the site of injury [14]. Trauma, infection, and pathogens that are physical and biological stimuli give innate response in form of inflammation and disrupt the homeostasis of organism. Long-term cell dysfunction results from chronic inflammation to restore normal immune function. Obesity, cancer, and liver and heart diseases are often involved in chronic inflammation. Betalains in beetroot supplementation presented anti-inflammatory shield in vivo [15].

Arachidonic acid is converted to leukotrienes and prostaglandins by lipoxygenase and cyclooxygenase (LOX and COX), the two bifunctional enzymes. Leukotrienes and prostaglandins are inflammatory chemical mediator. Betalainic effects against inflammatory cellular mediator are studied by several researchers [2]. The anti-inflammatory compounds present in beets have inhibited the activity of enzymes, which are involved to trigger inflammation such as cyclooxygenase [2]. But some inflammation is needed by the body to endure daily life tasks because it keeps us away from external bacterial or virus intrudement and from constantly getting sick. High level of inflammation is an alarm. Reduction in inflammation helps almost every system within the body to fight against diseases such as heart-related disease, diabetes, and disorder in the central nervous system (Alzheimer and Parkinson's diseases).

6.3 Regulatory Function of Phytochemicals

Community in which vegetal food is used as large portion of diet is less vulnerable to diseases such as cancer and atherosclerosis or CVD (disorders based on inflammation related to age) because these vegetables are beneficial due to the presence of secondary metabolites. These phytochemicals involve reducing properties which may effectively regulate the inflammation [14].

Atherosclerosis is a lasting inflammation of arterial intimae that progresses slowly during lifespan. In western world, atherosclerosis and heart-related diseases had been the leading basis of death [16]. Due to which, research on dietary compounds that prevent inflammation has growing interest. Indicaxanthin micromolar concentrations can regulate the expression of ICAM-1 as revealed by an in vitro inflammatory model containing endothelial cells of umbilical veins which were stimulated by inflammatory cytokine tumor necrosis factor-α (TNF-α) before the inflammation [16, 17].

The diet of a normal person contained some processed foods that are highly concentrated with inflammatory properties because it contains sufficient amount of sugar but low nutritional value. Therefore, taking beets is a superior way to diminish the immune system from constant harmful height of inflammation. Beets are gifted to combat inflammation because of betalain that is in fact activated inside the body from the nutrient choline and derived from vitamin B.

6.4 Cardiovascular Protective Effects

For cardiovascular health betalains are very important [18]. Betanin is reported to be effectively responsible for inhibition of lipid peroxidation [6]. In vitro studies revealed that betanin and betanidin are effective to stop lipid peroxidation of membranes and oxidation of low-density lipoproteins (LDLs). Betanin was highly effective inhibitor than betanidin as it could inhibit moderate percentage of lipid peroxidation in liposomal system. Both betanin and indicaxanthin could inhibit oxidative injury to LDL, whereas indicaxanthin was more effective due to its greater bioavailability. Betalain-concentrated erythrocytes were highly resistant toward membrane lipid peroxidation due to cumene hydroperoxide and subsequent hemolysis. Betanin also protects hydrogen peroxide-induced and nitrate-induced DNA damage aside from lipid peroxidation via RSA. LDL oxidation was also protected in vivo by betalains via transactivation of paraoxonase 1 (PON1) which was a hepatic antioxidant enzyme [2].

By oxidized LDLs, atherosclerosis is caused, as well as expression of monocyte chemoattractant proteins is induced minimally in vascular endothelial cells due to which monocytes are promoted in permeation and creation of foam cells in walls of coronary artery. It has been observed that the oxidative status of LDL became better by complementing healthy humans with fruits of Sicilian *Opuntia ficus-indica*, as decrease of present LDL hydroperoxides is proof of this. Moreover, plasma of healthy volunteers purified the LDLs after a single ingestion (500 g) of Sicilian *Opuntia ficus-indica* fruit pulp. Then after 3 h, they were more resistant to copper-induced oxidation than before the intake of the fruit meal [19].

Using indicaxanthin, ex vivo thwarting of human lifeblood plasma was done followed by isolation of LDL. Indication of betalain binding to LDL with an extreme binding of 0.5 nmol/mg in a saturable fashion was obtained. Toward oxidation induced by copper, indicaxanthin-enriched LDL is resistant as compared to homologous native particles, as estimated by increase in time required to initiate formation of lipo-peroxide [20]. In proteins like hemoglobin (Hb), myoglobin (Mb), and myeloperoxidase (MPO), the oxidation state of the heme iron should have been controlled so that the proteins could attain their appropriate practical activities, and not act as powerful oxidants. In the existence of organic hydroperoxides or hydrogen peroxide, these proteins were subjected to two-electron oxidation to produce iron radical species (X•[FeIV=O]) with greater oxidation state due to their peroxidase or peroxidase-like pyrrole [21–23]. If high-valent iron radical species was not rapidly reduced, it oxidized cell organelles resulting in the damage of proteins thus spoiling their role. The iron of hemeprotein is oxidized rapidly under uncontrolled conditions such as in β-thalassemia [24]. Continuous slow oxidation of its own takes place in healthy individuals causing red blood cells oxidative stress [25, 26].

Perferryl-Mb is a highly oxidizing hypervalent-iron myoglobin produced by digestion of meat [27]. Its radical scavenging causes preservation of oxidizable lipids and avoids production of potentially toxic lipid hydroperoxides. Concentrations of indicaxanthin consistent with a dietary intake were responsible for reducing perferryl-Mb as well as stopping the oxidation of lipids in heated red meat under a virtual gastric digestion. In the protection against attacking pathogens, MPO played key roles by oxygen-dependent antimicrobial activities [28, 29]. Myeloperoxidase (MPO) has harmful potential in the oncoming and advancement of atherogenetic processes and has considerable enzymatic tendency to oxidize LDL [30–33]. MPO may catalyze equally one- and two-electron oxidation reactions that are evident by their halogenation and peroxidase cycles [34], thus leading to production of cytotoxic hypochlorous acid and hypervalent iron redox intermediates compound II (CII) and compound I (CI). Indicaxanthin represents a good electron contributor for both CII and CI, at even very little μmolar concentrations. It was also suggested that betalain could scavenge hypochlorous acid [17, 35].

6.5 Anticancer Activities

The most important property of a pharmacoactive natural product is its tendency to act against cancer in the body. Due to wide spreading of complex diseases, these pharmacoactive natural products have drawn more attention [2]. Human chronic myeloid leukemia cell lines showed antiproliferative effects and apoptosis induced by betalains. Betalains, particularly betanin [7], are also powerful inducers of phase II purification enzymes, suggesting that they may help to eliminate xenobiotic-induced oxidative stress.

Formation of tumor is the main cause of cancer, which is related to inflammation due to oxidative stress and ROS (reactive oxygen species) generation. Few betalains including betacyanin have radical scavenging activity due to which they are important as anticancer [18]. Studies on different cell lines have discovered the dose-dependent inhibition of betalain pigments in cancer cell growth and spread [36].

In pure chemical form, betanin (which makes up 95% of the total betacyanins) is not as stable as in beetroot juice. In chemopreventive reduction of cell proliferation, inflammation, and esophageal, skin, lung, and liver cancer, *Beta vulgaris* (red beetroot) is a powerful agent. Diarrhea, constipation, hair loss, and vomiting are the side effects of DOX (a drug used by cancer patient) that can be eliminated by the use of beetroot juice which is safer and more cost-effective and improves life quality [18].

Betalains were investigated as a trans-active main transcription factor (Nrf2) that induced inner cellular antioxidant reluctant mechanisms. Supplementation of cells Huh7 with 15 mM of betalains via red beet caused a significant strengthening in cellular glutathione (GSH) level through which induction of cell period arrest and apoptosis by an important cytosolic antioxidant suggested the possible chemoprotective properties of betalain [37].

In cancer cells, *Opuntia ficus-indica* containing betalains effects as antiproliferative activity. Synthetic chemopreventive agent 4-HPR has inhibitory effect on growth of ovarian cancerous cells of human beings. Furthermore, these extracts suppressed the growth from 40% to 60% of well-kept cervical cancer cells and cervical epithelium cells. Betalains in actual betanins were found to have time and dose dependency in apoptosis of living chronic myeloid leukemia cell line (specifically K562) and also with immediate presence of cytochrome C in the cytosolic fractions of cell that decreased the membrane potential of mitochondria. Thus, a grouping of drug action with betalains that stops cells at various cell cycle checkpoints (S and/or G1 phases) could be expectant strategy to constrain tumor survival.

The practice of cancerous cell lines like MCF7 and K562 allowed the examination of potential anticancer drugs such as betalains in a controlled, simplified, and reproducible atmosphere. Future studies involving in vivo testing on living cancer cell lines possibly approximate the directed human cancer cells have more chance of victory in assessing the anticancer properties of betalains [11].

In animals, significant anticancer effect is shown by E162 (red beet color). When administered in drinking water at very low doses (between 25 and 78 mg/ml), it had steadily reduced the occurrence of tumors in the skin, lung, liver, colon, and esophagus in several models of laboratory animals. These results had stimulated further research in exploring the anticancer effects of red beetroot extract with the ultimate goal of establishing its efficacy in humans. Beetroot extract and its constituents are also used as dietary supplements in cancer prevention. This is based on the stability to manage oxidative stress involved in the origin and aggravation of cancer and the prevailing consensus that long-term daily exposure to small quantities of antioxidant dietary components has cancer chemopreventive potential. As beetroot extract has no known toxicity in humans, studies exploring its potential in mitigating toxic side effects of currently used potent anticancer drugs, such as doxorubicin (Adriamycin) when used in combination for synergistic chemotherapeutic effect, are also reviewed.

In all antitumor studies in laboratory, animals have been conducted exclusively with beetroot extract utilizing only two murine species: mouse (HOS:HR-1, ICR and SENCAR species) and rat (F344Harlan Sprague-Dawley and Wistar species). The drinking water containing red beetroot extract (red food color E162) exhibited significant anticancer activity by inhibiting the induction of lung and skin tumors in chemical carcinogen-treated mice. Encouraged by such observation, a systematic search was made for cancer chemopreventive agents among the natural colorants and related products of biologic and synthetic origin. Subsequent studies have established the multi-organ tumor chemopreventive effects of red beetroot extract in multiple species; the administration of red beetroot extract in drinking water showed no poisonous effect in any of the tested animals. There was no noticeable difference [1].

Betacyanin extract obtained from beetroot when showing to a compelling tumor promoter (like *N*-nitrosomethylbenzylamine) might hinder tumor frequency in rats. It had noticeably inhibited cell propagation, tumorigenesis, and angiogenesis in esophageal lesions, assigning its radical shooting and anti-inflammatory properties. In other animals, chemopreventive properties of betacyanin juice had also been detected in cancerous cells of the lung, liver, and skin. These discoveries in human and animal cell lines suggested beetroot supplementation embrace potential future approach that help and accomplish few of the indications of inflammation around cancer [15].

Antitumor effect of betanin for skin tumor had been presented by DMBA and reinvigorated by a chemical 12-*O*-tetradecanoylphorbol-13-acetate in rats. That was clear by the reduction in papilloma percentage after serving betanin in the amount of 2.5 mg/100 mL water for 25 weeks. They detected 40% shrinkage in the frequencies of 4-nitroquinoline-l-oxide (4-NQO) initiated and 8% glycerol-stimulated pulmonary tumors. Other than this, skin tumor is initiated through DMBA and stimulated by ultraviolet light, and related symptoms of splenomegaly were dramatically subdued through oral nurturing of betanin (concentration = 2.5 mg/100 mL water). Furthermore, betanin dose could considerably inhibit some other tumors immediately induced by different tumor originator and promoter application. Studies showed that there was decrease in incidence and proliferation of two-stage hepatocarcinogenesis which is a model initiated by the NDEA (initiator) and promoted by phenobarbital. On these facts, scientists suggested the probable use of betanin to combat against distortion. Anticancer activity had molecular mechanism seemed to be that betanin lessen the number of CD31+ endothelial microvessels to avoid angiogenesis as well as increases the appearance of caspase-3 to persuade apoptosis.

Previous studies had revealed antiproliferative properties of betanin for several cell lines with IC50 quantities ranging from 142 to 164 μg/mL. Betanin's cytotoxicity with IC50 (40 μmol/L) against human chronic myeloid leukemia cell line (specifically K562) was composed via cellular events like release of cytochrome C directly into the cytosol, poly (ADP-ribose) polymerase (PARP) breakdown, down-regulation of BCl-2, and reduction in mitochondrial membrane capacities.

Betanin had produced activation of motivator caspase-9 and effector caspase-3 and caspase-7 and breakage of caspase-3 target PARP as established in human lung cancerous cell lines. Correspondingly, isobetanin or betanin treatment focused on the stimulation of both extrinsic and intrinsic apoptotic routes in cell line of breast cancer (specifically MCF-7). This experiment is performed on mouse cells, myeloid leukemia cultures, lung cancer cell lines, and breast cancer cell line that suggested the anticancer properties of betanin by process of apoptosis. In another model of activated neutrophils, a portion of anti-inflammatory reaction, betanin had expressively boosted the properties of caspase-3 and its breakdown products, though remaining neutrophils had not any outcome. Dietary indicaxanthin could pointedly avoid expiry of erythrocyte- or eryptosis-linked events and atherosclerotic problems produced by cytotoxic oxysterols [2].

6.6 Effects on Blood

It was exposed by one of the preliminary studies on bioactivity of betalains that red beet betanin when inserted caused brief escalation of heart rate and blood pressure [2]. Red blood cells (RBCs) are readily capable to oxidation due to high oxygen tension, high iron amounts, and transition metals that favor oxygen free radical production [39]. These oxidized RBC's cause cell function deterioration and untimely aging and endothelial cells injury [26]. Dietary indicaxanthin produces protective effects. β-Thalassemia, a genetic hemolytic disorder, is caused by high production of reactive oxygen species. In order to decrease RBC antioxidant defense that causes cell organelle damage, morphology deterioration and cell membrane dysfunction pathology are helpful [24, 40–43]. In β-thalassemia patients, indicaxanthin's protective dose-dependent effects were studied in RBC when subjected to in vitro oxidation by cumene hydroperoxide [38]. Betalain arriving the RBCs had high resistance of thalassemia RBC to hemolysis and prevented Hb oxidation and lipid that retarded the GSH exhaustion and vitamin E [17].

6.7 Antimicrobial Activity

Betalains possess antimalarial and antimicrobial activities. Noteworthy antimalarial properties in mice were because of high level of betanin. In betalain-rich *Amaranthus spinosus*, vital inner cations (Fe^{2+}, Ca^{2+}, and Mg^{2+}) were chelated by amaranthin due to which intracellular transport of parasites were blocked. Growth rate of *Escherichia coli* had been inhibited by the extracts of *Opuntia matudae* containing betalains. Zones of reduced growth in *Salmonella typhimurium* are induced by beetroot *pomace*, *Staphylococcus aureus*, and *Bacillus cereus*. Gram-negative bacteria (*E. coli*, *Pseudomonas aeruginosa*, *Citrobacter freundii*, *Citrobacter youngae*, *E. cloacae*, *Salmonella typhimurium*) with *S. typhimurium* and *C. freundii* were inhibited by beetroot *pomace* which is most influenced. A broad antimicrobial spectrum is applied by betalain-concentrated extracts obtained from red pitahaya through hindering some Gram-positive bacteria (*B. cereus*, *S. aureus*, *Escherichia faecalis*, and *Listeria monocytogenes*) at 7.8 mg/mL. Just because of their negative effects on the structural, functional, and cellular membrane penetrability of the microbes, they exhibit antimicrobial activity of betalains which finally leads to cell death. A wide antimicrobial spectrum is inhibited by betalains, on their mechanism of microbial inhibition very little of these are describe in the literature. In future, we should investigate the specific basic cellular and molecular mechanism of antimicrobial activity of betalains [11].

6.8 Diuretic Effect

By using dehydrated *Opuntia ficus-indica* fruit extract containing high contents of reducing sugars, diuretic effects of betalains were studied compared to that of hydrochlorothiazide, which is a standard drug [2].

6.9 Pain-Relieving Effects

By using red beet extract, original betalain was formed that contained importantly high (24.6%) betalain content and tends to moderately relieve osteoarthritis pain feeling in human volunteers. The discoverer proposed that this formulation will also work in several conditions such as acne, contact dermatitis, sinusitis, and allergy. However, open-type clinical experiments were performed instead of clinical efficacy study, since only 1/1 male and female volunteer was subjected for the acne-related study [2].

6.10 Hypoglycemic Effect

Betalains were able to antagonize diabetic complications in conditions of chronic hyperglycemia. A report was published in 2012 on the hypoglycemic effect of betanidin in rats. Sugar level of blood was reduced 50% in rats that were injected with diet of (24 weeks) or complemented with betanidin (9.6 mg) for 40 days. This study sets off a few other investigations on hypoglycemic effect of betalain-rich preparations or purifications [2].

6.11 Hepatoprotective Effect

Quinone reductase, a phase II enzyme, was extracted from red beet, and induction was carried out in order to observe the hepatoprotective effect of betalain especially betanin. Induction in rats showed that hepatic toxicity produced by carbon tetrachloride, 7,12-dimethylbenz(a)anthracene (DMBA), *N*-nitrosodiethylamine (NDEA) [2], lactate dehydrogenase, aspartate aminotransferase, gamma-glutamyl transferase, and alanine aminotransferase [6] can be relieved by red beet juice containing high concentrations of betalains due to improved antioxidant levels in the body. Toxicity of the liver was cancelled expeditiously by implying betanin's protective effects because it produced an improved redox status of the liver and restores mitochondrial activities [2].

6.12 Radioprotective Effect

Betalains have antiviral and radioprotective activity which is suggested by their production from ice plant (*Mesembryanthemum crystallinum L.*) by inducing UV radiation and from red beet (*Beta vulgaris L.*). Red beet extract was used in order to check the radioprotective efficiency of betalains. It was noted the betalains were able to improve the count of white blood cells, antioxidative strength of the body, and micronuclei in polychromatophilic erythrocytes of bone marrow, spleen, and thymus index in mice treated by ^{60}Co [2, 7].

6.13 Neuroprotective Effects

Neurotoxicity is caused by substances such as d-galactose, and it was diagnosed by oxidative stress in the brain of matured rates. Studies revealed the neuroprotective competence of betacyanin as compared to vitamin C. Due to which it was able to delay the symbols of aging in brain tissues to some degree [2].

6.14 Gene Regulatory Activity

Betanin as a killer of oxygen radicals displayed gene regulatory properties partially by nuclear factor (specifically erythroid-derived 2) like 2-(Nrf2) focusing on signaling routes. Betanin could also prompt phase II enzymes as well as antioxidant resistance system [15].

6.15 Toxicity Activities

Toxicology refers the effect of harmful chemicals to the environment. The toxic effect of colors and pigments in food cannot be ignored due to health concerns. According to the report of regulatory authorities of the USA, kids are consuming more colored foods in actual than expected, as the quantity has increased from 12 to 62 mg/capita/day from 1950 till 2010. Synthetic azo dyes are the major source of food additives which impart color, and their toxicity is gaining high attention. These are very popular due to inexpensiveness and attractive bright colors.

The azo dyes such as Sunset Yellow, Tritrazine, Azorubine, Allura Red, and Patent Blue are able to bond strongly with human serum albumin and produce severe toxins. As new discoveries happen in the world of dyes like blue dyes are utilized to excise and visualize a central lymph node in cancer patients. Some side reactions including hypersensitivity have become more frequent. Clinical studies reveal that these colorant mixtures when consume have different effects on the body.

The main concern is to limit the utility of these dyes which are produced by intestinal microbiota due to azo reduction and cause carcinogenicity. Although these metabolites are created in the human body, the medical effects of these dyes are subjected to the amount of colorant consumed. However, in light of new discoveries, it is essential for the regulatory authorities to evaluate the possible toxicity of food colorants and thus revise instructions for their use. In 2008, the European Parliament and the Council (EC) has published Regulation No., that is, 1333/2008 on edible food additives.

Betalain has been used as food colorant for many common dairy products like ice cream, yogurt, powdered drink, cake mix, frostings, soft drinks, candies, gravy mixes, meat substitutes, marshmallow candies, and gelatin deserts. It is presumed that less than 50 mg/kg of betanin has produced desired color; however, higher quantity has also been reported in literature [44].

Dietary safety Safety is the prime factor for dietary consumption by living organism. In initial reports by a FAO/WHO Committee, it was showed that there was no proper evidence on metabolism and safety of beetroot pigments. Afterward, thorough studies on metabolism absorption, cardiovascular effects, and excretion of beetroot extract were accompanied. It was found that in oral administration of betanin, the main pigment in red beet was not absorbed effectively and the major portion was metabolized in the gut. Moreover, betanin quickly enhances blood pressure and heart rate. Also, beet pigments never indorsed hepatocarcinogenesis and betanin inhibited IgE and IgG production which suggest lack of allergic response by the pigments. Other aspects of protection such as mutagenicity, absence of genotoxicity, and short-term toxicity of beet betalains on *Salmonella typhimurium* and rats were also documented. Recently, no mutagenicity in *S. typhimurium* is observed after exposure to betalain-rich extracts obtained by fruits of some *Cactaceae* species. It concluded that betalains are safer food items. Although the work and observations on embryo toxicity, including generation toxicity and teratogenicity, are not well established, one aspect that could be linked to the utility of beet pigments was beeturia (the phenomenon which is based on excretion of colored urine after consuming red beet).

There are many variable reports on beeturia. It was properly observed and documented that beeturia was not a property of an individual physiological symptom and had no prove of its existence under polymorphic genetic control. The recent literature indicated that beeturia was an indication to measure consumption and co-ingestion when incorporated with organic acids such as oxalic acid, ascorbic acid, and flow of gastric emptiness. Beeturia was only inferred physiological disorder caused by excessive intake of betalain-rich food and not associated with any kind of disease. Other than profound interest in the functionality of betalain sources, such as quinoa and amaranth, most of the research of betalains to attain biological activity, bioavailability, and dietary safety have been conducted on red beetroot. It is a source of pigments due to its easy availability and bioaccessiblity, so red beet betanin is a signified red food colorant. This is the reason that betanin is recognized as food colorant that is red by the European Union of Food and Drug Administration (FDA), USA.

In current studies, several new sources of diverse betalains have been documented. Some of which are considered dangerous for health. For example, betalain obtained from *Phytolacca decandra* is one of the promising barriers that holds toxic saponins due to which these berries are not recommended commercially. In *Celosia argentea* var. *cristata*, owing to have increased level of dopamine (41.15 μmol/g) when fresh, there is a chance of ingesting from this plant of betalain-rich extract.

A large number of structurally modified and varied betalains from plant sources were reported. Toxicological studies of betalain-rich contents from *Myrtillocactus geometrizans*, *Rivina humilis*, and *Hylocereus polyrhizus* fruits have been performed on rodents and revealed that these extracts have nontoxic effect on human health. Moreover, betalain extract obtained from hairy roots of red beet are nontoxic for rats, and so they suggest that safe betalains are produced through biotechnology that has sustainability.

Standard method of pigmentation is providing temperature, while prolong period of drying at normal temperature may enhance yellow components of color. To overwhelm these problems, some known techniques employed for betalain concentration include osmotic distillation, fermentation, and convective drying. Fermentation to concentrate betalain from *Opuntia stricta* fruit and the pigment having filtrate was found very safe. Following the same guidelines, lacto-fermentation is used to concentrate betalains getting from roots of *Beta vulgaris* [2].

For the safety of commercially processed betalains other than conventional methods of extraction, they have been positively extracted with pulsed electric fields (PEF). It induces tissue damage while permeability allows the release of pigments. Similarly, the techniques such as microwave, enzyme treatment coupled with microwave, aqueous two-phase technique, γ-radiation, enzyme treatment, pressurized carbon dioxide, and thermo-ultrasonication and some other novel methods have been utilized for betalain extraction. Additionally, betalains extracted after enzyme pretreatment method and pulsed electric fields were observed to be friendly and safe. Other than these, no evidence on bioavailability, biological activity, and safety of novel betalains has been documented via novel methods.

Because of the enhanced mutual extraction of antioxidants like phenols and flavonoids, few of these processing methods have been cited as favorable in enhancing antioxidant power or bioactivity of plant-extracted betalains. Moreover studies showed that beetroot juice has high concentration of neobetanin (degradation parts of betanin produced during processing) that decreased the postprandial insulin and glucose response.

On the other side, some of the highly used processing techniques for the production of betalain concentrates are freeze-drying, spray-drying, and/or air-drying that produce variety in pigment composition and other phytochemicals in the end product. In the reality of the above studies, a huge scientific discussion on the stability, applicability, and safety of betalain-rich contents synthesized during the various processing methods is required. In a nut shell, this is due to conflicting claims of low stability, incomplete pigment extraction, and low pigment content related to handling methods. Therefore, in non-lacto-fermentation process of betalain hit extract, flavor, and safety aspects, while in case of fermentation, final products and undesired by-products need to be studied in a proper way.

References

1. Chen, C. (Ed.). (2015). *Pigments in fruits and vegetables: Genomics and dietetics* (pp. 127–140). New York: Springer.
2. Khan, M. I. (2016). Plant betalains: Safety, antioxidant activity, clinical efficacy, and bioavailability. *Comprehensive Reviews in Food Science and Food Safety, 15*(2), 316–330.
3. Esatbeyoglu, T., Wagner, A., Schini-Kerth, V. B., & Rimbach, G. (2015). Betanin – A food colorant with biological activity. *Molecular Nutrition & Food Research, 59*, 36.
4. Mohamed Yahya, K. (2006). *Studies on tissue culture system for the production of food Colours from Beta Vulgaris L.* Doctoral dissertation, University of Mysore.
5. Kanner, J., Harel, S., & Granit, R. (2001). Betalains; a new class of dietary Cationized antioxidants. *Journal of Agricultural and Food Chemistry, 49*, 5178–5185.
6. Clifford, T., Howatson, G., West, D. J., & Stevenson, E. J. (2015). The potential benefits of red beetroot supplementation in health and disease. *Nutrients, 7*(4), 2801–2822.
7. Tanaka, Y., Sasaki, N., & Ohmiya, A. (2008). Biosynthesis of plant pigments: Anthocyanins, betalains and carotenoids. *The Plant Journal, 54*, 733–749.
8. Kathiravan, T., Nadanasabapathi, S., & Kumar, R. (2014). Standardization of process condition in batch thermal pasteurization and its effect on antioxidant, pigment and microbial inactivation of Ready to Drink (RTD) beetroot (Beta vulgaris L.) juice. *International Food Research Journal, 21*(4), 1305–1312.
9. Czapski, J., Mikołajczyk, K., & Kaczmarek, M. (2009). Relationship between antioxidant capacity of red beet juice and contents of its betalain pigments. *Polish Journal of Food and Nutrition Sciences, 59*(2), 119–122.
10. Kugler, F., Stintzing, F. C., & Carle, R. (2007). Evaluation of the antioxidant capacity of betalainic fruits and vegetables. *Journal of Applied Botany and Food Quality, 81*(1), 69–76.
11. Gengatharan, A., Dykes, G. A., & Choo, W. S. (2015). Betalains: Natural plant pigments with potential application in functional foods. *LWT-Food Science and Technology, 64*(2), 645–649.

12. Ravichandran, K., Saw, N. M. M. T., Mohdaly, A. A., Gabr, A. M., Kastell, A., Riedel, H., et al. (2013). Impact of processing of red beet on betalain content and antioxidant activity. *Food Research International, 50*(2), 670–675.
13. Gliszczyńska-Świgło, A., Szymusiak, H., & Malinowska, P. (2006). Betanin, the main pigment of red beet: Molecular origin of its exceptionally high free radical-scavenging activity. *Food Additives and Contaminants, 23*(11), 1079–1087.
14. Allegra, M., Ianaro, A., Tersigni, M., Panza, E., Tesoriere, L., & Livrea, M. A. (2014). Indicaxanthin from cactus pear fruit exerts anti-inflammatory effects in carrageenin-induced rat pleurisy. *The Journal of Nutrition, 144*(2), 185–192.
15. Esatbeyoglu, T., Wagner, E. A., Schini-Kerth, V. B., & Rimbach, G. (2014). Betanin- A food colorant with biological activity. *Molecular Nutrient Food Research, 59*, 36–47.
16. Gentile, C., Tesoriere, L., Allegra, M., Livrea, M. A., & D'alessio, P. (2004). Antioxidant Betalains from Cactus Pear (Opuntia ficus-indica) Inhibit Endothelial ICAM-1 *Expression. Annals of the New York Academy of Sciences, 1028*(1), 481–486.
17. Livrea, M. A., & Tesoriere, L. (2015). Indicaxanthin dietetics: Past, present, and future. In *Pigments in Fruits and Vegetables* (pp. 141–163). New York: Springer.
18. Das, S., Williams, D. S., Das, A., & Kukreja, R. C. (2013). Beet root juice promotes apoptosis in oncogenic MDA-MB-231 cells while protecting cardiomyocytes under doxorubicin treatment. *Journal of Experimental Secondary Science, 2*, 1–6.
19. Livrea, M. A., & Tesoriere, L. (2006). Health benefits and bioactive components of the fruits from Opuntia ficus-indica [L.] Mill. *Journal of the Professional Association for cactus Development, 8*(1), 73–90.
20. Tesoriere, L., Butera, D., D'arpa, D., Di Gaudio, F., Allegra, M., Gentile, C., & Livrea, M. A. (2003). Increased resistance to oxidation of betalain-enriched human low density lipoproteins. *Free Radical Research, 37*(6), 689–696.
21. Everse, J., & Hsia, N. (1997). The toxicities of native and modified hemoglobins. *Free Radical Biology and Medicine, 22*(6), 1075–1099.
22. Everse, J. (1998). The structure of heme proteins compounds I and II: Some misconceptions. *Free Radical Biology and Medicine, 24*(7), 1338–1346.
23. Furtmüller, P. G., Obinger, C., Hsuanyu, Y., & Dunford, H. B. (2000). Mechanism of reaction of myeloperoxidase with hydrogen peroxide and chloride ion. *The FEBS Journal, 267*(19), 5858–5864.
24. Rund, D., & Rachmilewitz, E. (2005). β-Thalassemia. *New England Journal of Medicine, 353*(11), 1135–1146.
25. Vollaard, N. B., Reeder, B. J., Shearman, J. P., Menu, P., Wilson, M. T., & Cooper, C. E. (2005). A new sensitive assay reveals that hemoglobin is oxidatively modified in vivo. *Free Radical Biology and Medicine, 39*(9), 1216–1228.
26. Rifkind, J. M., & Nagababu, E. (2013). Hemoglobin redox reactions and red blood cell aging. *Antioxidants & Redox Signaling, 18*(17), 2274–2283.
27. Halliwell, B., Zhao, K., & Whiteman, M. (2000). The gastrointestinal tract: A major site of antioxidant action? *Free Radical Research, 33*(6), 819–830.
28. Shah, S. V. (1989). Role of reactive oxygen metabolites in experimental glomerular disease. *Kidney International, 35*(5), 1093–1106.
29. Klebanoff, S. J. (1975). Antimicrobial mechanisms in neutrophilic polymorphonuclear leukocytes. *Seminars in Hematology, 12*(2), 117–142 Elsevier.
30. Hazell, L. J., Arnold, L., Flowers, D., Waeg, G., Malle, E., & Stocker, R. (1996). Presence of hypochlorite-modified proteins in human atherosclerotic lesions. *Journal of Clinical Investigation, 97*(6), 1535.
31. Hazell, L. J., & Stocker, R. (1993). Oxidation of low-density lipoprotein with hypochlorite causes transformation of the lipoprotein into a high-uptake form for macrophages. *Biochemical Journal, 290*(1), 165–172.
32. Hazen, S. L., & Heinecke, J. W. (1997). 3-Chlorotyrosine, a specific marker of myeloperoxidase-catalyzed oxidation, is markedly elevated in low density lipoprotein isolated from human atherosclerotic intima. *The Journal of Clinical Investigation, 99*(9), 2075.

33. Kostyuk, V. A., Kraemer, T., Sies, H., & Schewe, T. (2003). Myeloperoxidase/nitrite-mediated lipid peroxidation of low-density lipoprotein as modulated by flavonoids. *FEBS Letters, 537*(1–3), 146–150.
34. Dunford, H. B. (1999). *Heme peroxidases*. New York/Toronto: Wiley-vch.
35. Allegra, M., Furtmüller, P. G., Jantschko, W., Zederbauer, M., Tesoriere, L., Livrea, M. A., & Obinger, C. (2005). Mechanism of interaction of betanin and indicaxanthin with human myeloperoxidase and hypochlorous acid. *Biochemical and Biophysical Research Communications, 332*(3), 837–844.
36. Gandia-Herrero, F., & Garcia-Carmona, F. (2013). Biosynthesis of betalains: Yellow and violet plant pigments. *Trends in Plant Science, 18*(6), 334–343.
37. Davies, K. (Ed.). (2009). *Annual plant reviews, plant pigments and their manipulation* (Vol. 14). New York: John Wiley & Sons.
38. Tesoriere, L., Allegra, M., Butera, D., Gentile, C., & Livrea, M. A. (2006). Cytoprotective effects of the antioxidant phytochemical indicaxanthin in β-thalassemia red blood cells. *Free Radical Research, 40*(7), 753–761.
39. Halliwell, B., & Gutteridge, J. M. (2015). *Free radicals in biology and medicine*. New York: Oxford University Press.
40. Chiu, D. T. Y., Van Den Berg, J., Kuypers, F. A., Hung, I. J., Wei, J. S., & Liu, T. Z. (1996). Correlation of membrane lipid peroxidation with oxidation of hemoglobin variants: possibly related to the rates of hemin release. *Free Radical Biology and Medicine, 21*(1), 89–95.
41. Grinberg, L. N., Rachmilewitz, E. A., Kitrossky, N., & Chevion, M. (1995). Hydroxyl radical generation in β-thalassemic red blood cells. *Free Radical Biology and Medicine, 18*(3), 611–615.
42. Scott, M. D., Van den Berg, J. J., Repka, T., Rouyer-Fessard, P., Hebbel, R. P., Beuzard, Y., & Lubin, B. H. (1993). Effect of excess alpha-hemoglobin chains on cellular and membrane oxidation in model beta-thalassemic erythrocytes. *Journal of Clinical Investigation, 91*(4), 1706.
43. Van Dyke, B. R., & Saltman, P. (1996). Hemoglobin: A mechanism for the generation of hydroxyl radicals. *Free Radical Biology and Medicine, 20*(7), 985–989.
44. Amchova, P., Kotolova, H., & Ruda-Kucerova, J. (2015). Health safety issues of synthetic food colorants. *Regulatory Toxicology and Pharmacology, 73*(3), 914–922.

Chapter 7
Betalains as Colorants and Pigments

Colors are the unique identification of any object. A colorant can act as either a pigment or a dye. Pigments are the chemicals that absorb light of specific wavelength and prevent certain wavelengths of light from being transmitted or reflected. They are also regarded as subtractive colors. The colors of pigments interpret in the brain by producing neural impulse when eye seizures the diverted or reflected non-engrossed energy.

Both synthetic and natural pigments are exploited in food, cosmetics, clothes, medicines, and many supplementary life accessories. Similarly, chlorophyll is a pigment which support in photosynthesis, being the primary function of food preparation in plants. Furthermore, pigmentation is also used in sending signals for courtship and reproductive behavior in animals as well as protecting them [1]. The intemperance or moderation in pigment development in humans and animals resulted in abnormal conditions and range of diseases. Plants are considered as primary source of colors, dyes, and pigments. Being natural, they are safe and impart positive health effects.

The chemical compounds which reflect only certain wavelength of visible light are called pigments, and they are considered as colorful substances. In plants, electronic structure of the pigment is responsible for pigmentation when it interacts with sunlight and modifies the wavelengths which is either transmitted or returned by the plant tissue. Electrons jump from lower orbital to higher orbital by absorbing energy, this precise part of molecule is known as chromophore.

Pigments are usually insoluble dyes. They are insoluble in common solvents exhibiting good cover and color properties. In organic pigments such as calcium carbonate, titanium dioxide, adsorbent carbon, and red iron oxides can be employed with a limited variety of colors. Organic pigments are typically in the form of lacquers which are insoluble complex salts of water-soluble azo dyes in a wide color palette [1].Wavelengths between 380 and 730 nm are usually detected by humans without color blindness, demonstrating the visible spectrum of red, orange, yellow, green, blue, indigo, and violet. Green color appears when peak absorbencies of chlorophyll are at 430 and 680 nm. Usually, the colors are the product of a mixture

E. Akbar Hussain et al., *Betalains: Biomolecular Aspects*,
https://doi.org/10.1007/978-3-319-95624-4_7

of residual wavelengths; for instance, yellow-green light has a wavelength of 520–530 nm and is captivated by anthocyanins which will create mauve colors made by the likeness of a mixture of orange, red, and blue wavelengths.

7.1 Some Significant Plant Pigments

Chlorophyll Chlorophylls are the most evident and prevalent pigments of plants. These are magnesium chelated with cyclic tetrapyrrole pigments and bear a resemblance to the hem portion of hemoglobin and bile pigments of animals responsible for photosynthesis.

Carotenoids The pigments creating vivacious colors to fruits and flowers are due to the carotenoids. They are terpene-based pigments present in all photosynthetic plants and microorganisms such as *Erwinia* and *Rhodobacter*, are essential components of photosystems, and are responsible for yellow to red colors of flowers and fruits. Carotenoids along with flavonoids and anthocyanins usually occur in the same vegetative organs, and their combination increases color variety.

Betalains as Natural Pigments Betalains are the major group of pigments that occur naturally, are nitrogen-based molecules, and are found only in Order *Caryophyllales* and limited number of some fungi families. They are the most taxonomically limited plant pigments. Being natural pigments, they are more stable than synthetic dyes and do not cause health-damaging effects. Inquisitively, their incidence is reciprocal to that of anthocyanins. Colors induced by betalains are responsible for attracting animals toward flowers for pollination. Due to visible light emission, an additional signal recommended in plant pollinators for a short time is engrossed as a result of the conservation of visible fluorescence in betaxanthins. Due to the presence of betaxanthins, flowers are yellow, while betacyanins impart bright violet color to the plants. By removing oxygen, betalain stability is amplified with the usage of ascorbic and isoascorbic acids as antioxidants.

7.2 Betalains as Food Colorants

Any pigment or dye that functions as color and used as additive that induces attraction and enhances the food presentation is called food colorant. This is not restricted to food only, but other non-food substances including pharmaceuticals use colorants. Moreover, any chemical substance when blended with an additional material and yields a new color is regarded as a color additive.

Soluble colorants are classified into synthetic, semisynthetic, and natural ones. Numerous food material or other natural materials combine to form natural dyes. They comprise of riboflavin (E 101), carotenes (E160a), chlorophylls (E140), anthocyanins (E 163), or betalain (E 162). Chemical synthesis has been carried out for dyes that do not occur naturally. Initially, these dyes were isolated from coal tar and nowadays are obtained from extremely cleansed oil products [2]. The organic synthetic colorants involved azo, anthraquinone, chinilin, and xanthane dyes, which usually have more intensive and enduring color range than natural substances and do not create any flavor to foodstuffs, and they are usually stable.

Betalains are used as food colorants since the 20th century. Pokeberry juice comprising of betalains has been used to improve the color of red wine and color meat and sausages, ice cream, and powdered soft drinks; other uses are in some sugar sweets, e.g., fondants, sugar coatings, sugar strands, and fruit or cream fillings. Moreover, it is used in soups along with tomato and bacon products. It can be added in hot processed candies during the final stage of processing. Betalains comprising extracts of red beets are used as natural food colorants and in pharmaceuticals and cosmetics. Betalain color is pH independent as compared to anthocyanins, so it is beneficial as compared to betalains. These pigments are unchanged at pH ranging from 3 to 7 and consequently good for use as colorant where taste may vary from sour to neutral in foods, while the use of anthocyanins is not conceivable at pH over 3 due to instability. Excluding the preparation of the natural colorant E162 from red beet, the extraction of E162 from cactus pear (*Opuntia*) and *Amaranthaceae* is also probable. Thus, in ice creams or yogurt with a low pH value, cactus pears can be used in foods.

Due to deficiency of betalain constancy in food mediums like dairy products or soft drinks, additional steps for better implementation of betalains as food colorants should be directed. Betalains from red beet (*Beta vulgaris*) is independent of pH and is more stable than that from anthocyanins.

Betalains obtained from pitahaya juice are better than red beet juice because of its higher pigment stability and high oxalate, nitrate, and earthbound microorganisms (the chief difficulties with red beet). Aqueous extracts of red beet available as juice or powders are limited around the world by regulation commercially. The juice concentrates are beet pigments (vacuum drying up to 60–65% total solids), classified as vegetable juices by Food and Drug Administration (FDA), to obtain a beet powder. They are commonly spray-dried with maltodextrin, 75–80% sugar, 10% protein and 0.3–1% betalains are found in them. Fifty percent of betalains can be recovered through the traditional way of getting beet juice such as hydraulic pressing; this process had been amplified by means of steeping enzymes. High variety of colors, odors, and flavors are available that are symmetrical to beet, and compensation is the least problem related with red beet extracts. Slight amounts of unadulterated pigments (<50 ppm calculated as betanin) are used to reach the wanted sort for most applications, as betalains have high molar absorptivity.

Free betalamic acid and betanidin occur in low amounts in betalamic plants. Since the molar extinction coefficient of betalamic acid (betalamic acid = 25,000 $Lmol^{-1}$ cm^{-1} 20) is relatively small, its potential color effect is small when betaxanthins (£mean = 48,000 L*mol”1 cm”1; 24) and betacyanins (mean = 60,000 mol’1 cm”1; 25) are present at the same time. Although betanidin (betanidin =49,000 $Lmol^{-1}$ cm^{-1}) exhibits the same extinction coefficient as betanin, it looks to be present in slight amount in plants (P). In addition to individual structures, the overall appearance of betalamic tissues will be governed by the total color content with a darkening effect through higher pigment contents, i.e., a yellow shade will appear orange if more concentrated, the same being true for red turning to a purple tone.

7.3 Structure of Betalains as Food Colorant

Betanin represents the most noteworthy and abundant pigment among the betalains. Betanin is a glycoside while betanidin is an aglycone. Betalamic acid's ammonium derivatives are water-soluble compounds. The protonated form of 1,2,4,7,7-pentasu bstituted-1,7-diazaheptamethin (yellow color) is chromophore of betalains. Three conjugated double bonds are responsible for color. Betalains have two groups depending on the ligand, yellow betaxanthins (480 nm) and the red-violet betacyanins (540 nm). Betanidin is a condensation product of betalamic acid (406 nm) and cyclo-3,4-dihydroxyphenylalanine (cyclo-DOPA). The structure assortment of betacyanins is vast due to the joining of the hydroxyl group at positions 5 and 6 of betanidin with glycosides or acyl glycosides. 5-*O*-glucosylated (i.e., betanin) linkage of betacyanins is more than 6-*O*-glucosylated (i.e., gomphrenin II); both positions are never glycosylated. Betacyanins are divided into four groups which are gomphrenin, descarboxybetanin, betanin, and amaranthin.

Betanin is structurally a glyconebetanidin that involves β-glycosidic linkages with a glucose unit at C5. Biogenic amines condense with betalamic acid and produce betaxanthins, vulgaxanthin in cactus pear, indicaxanthin from *Opuntia ficus-indica* (L), and yellow beet from *Beta vulgaris* (L). Betalains exist as isomers because of the chiral C-atom of the dihydropyridine unit. Factors affecting the betanin-containing content of red beet are farming conditions, temperature during the growing season, cultivar, soil moisture, soil fertility, and storage temperature. Different substitution patterns of betacyanins in plants containing betalain produce red and violet color. Betanidin glycosylation usually takes place with a hypsochromic shift of about 6 nm. Visual presence was not affected, while glycosylation at C-6 was considered to create a larger bathochromic shift than 5-glycosylation. Bathochromic shift of up to 12 nm have been seen with hydroxycinnamic acids of the esterification of betanidin 5-*O*- or 6-*O*-glycosides.

In contrast, aliphatic acyl moieties such as 3-hydroxy-3-methyl-glutaric acid in hylocerenin malonic acid in phyllocactin did not alter or slightly (2–3 nm) alter the wavelength. Color properties of betacyanins do not change by isomerization at C-15. Under acidic conditions, epimerization is decreased, when betacyanins were acylated with hydroxycinnamoylresidues, the bathochromic shift, and betacyanin stability increased through intramolecular stacking by substitution with these aromatic acids.

Yellow part of betacyanins is betaxanthins. All kinds of color shades are achieved by the plants through mixtures. Hypso- and bathochromic shifts are produced by condensation reaction of betalamic acid and various amino acids or amines. Lower absorption maximum is usually shown by the amine conjugates than their own amino acid counterparts, and indicaxanthin together with portulacaxanthin I exhibits the highest [3].

Yellow vulgaxanthin I is an additional core compound. Bicyclic alcohol geosmin ($C_{12}H_{22}O$ trans-1,10-dimethyl-trans-(9)-decalol) is produced by *Streptomyces* species and *Mycobacteria*, and the simple-stuffy aroma and flavor of red beets are because of the many 3-alkyl-2-methoxy pyrazines (3-sec-butyl-,3-isobutyl- and 3-isopropyl-2-methoxypyrazine). High amount of nitrate is found in red beet, salad, radish, rocket, and spinach. Nitrate contents are eliminated in juices by ionic exchangers or microbiological methods. Microbiological denitrification of red beetroot and betalain-containing foods is easily made by wild *Celosia argentea*, gram-negative bacteria *Paracoccus (Blitum capitatum, syn. Chenopodium capitatum), Ullucus tuberosus, Rivina humilis* L. berries, *Mammillaria* sp., and *Talinum triangulare* (Jacq.) [4].

7.4 Genetic Model of Betalain Pigmentation

A genetic model of betalain pigmentation on the exploration of cross inbred of *P. grandiflora* species with respect to betalain biosynthesis has been published. The subsequent segregation patterns show that in petal pigmentation, a minimum of three loci, R (red), C (color), and I (inhibitory), are involved.

The 3,4-dihydroxyphenylalanine DOPA is converted to betalamic acid due to the appearance of the dominant C locus which is responsible for the colored petals and shoots of the plants. Accumulation of betaxanthins is inhibited by dominant I locus. Fabrication of violet betacyanins is due to dominant R connected with the accessibility of cyclo-DOPA or its glycosylated form; *Portulaca* flowers are either pale yellow in the absence of the dominant R locus or deep yellow in presence of the dominant I with the homozygous recessive II background.

Fabrication of pure violet manner occurs due to the presence of dominant I locus; otherwise, orange-red color will be exhibited by flowers caused by the huge concentration of yellow betaxanthin over a violet betacyanin background. Loci I and R are sturdily associated at 5.3 cm. imino residues conjugation which do not contain cyclo-DOPA) or amino acid with betalamic acid is prevented by lessening the formation of betaxanthins and accrual of betacyanins in the violet phenotype (dominant R locus) increased. This inferred that the chromophore is related to the vacuole that is present in the cytoplasm for the rapid conjugation of betaxanthins. The existence of a specific transporter for betalamic acid would be partly or completely blackout in existence of the product of locus I. In spite of the manifestation of the inhibitor I in the pale yellow CCrrII phenotype, a small amount of yellow pigment is constantly formed. In such plants, free betalamic acid is produced in great amount; besides traces of dopaxanthin, no miraxanthin V (dopamine) has been noticed.

The presence of the products of the R and I loci is still unidentified and will need additional inquiries into controlling gene named as blotchy (bl) or uneven pigment modeling. Linkage between the R and Y loci and the bl gene is found, and two-point linkage between the R and Y loci was expected at 7.4 cm. The R-Y-bl genetic region is thus imperative in the genomic control in *B. vulgaris* for biosynthesis of betalain. The bl gene displayed some transmission alteration thus suggesting an epigenetic control among others [5].

7.5 Beetroot Waste as Source of Dyes

Beetroot waste is used as source of dyes for wool and polyamide dyeing before it is recycled and formed in the food industry either as a crumpled mass from the abstraction of beetroot juice or as a peel from beetroot sliced in salads. The yellow betaxanthins showed bands at 476 nm, and red betacyanins showed absorption band at 534 nm. In a strong alkaline medium, remarkable alteration in the shape of band can be seen, when the peak for the yellow dye constituent starts to decline.

Different resemblance has shown by beetroot dyes to the wool which was dyed red or orange by betacyanins, while betaxanthins polyamide was dyed only yellow. This can be used for streaked and other effects using mixed fabric or two component knitted fabrics [6]. A clear difference can be seen between the wool, dyed in the neutral and acidic environment to shade from orange to reddish purple, and yellow color shades can be found in polyamide fabric. Pre-treatment using alumim treated colors and pre-treated woolen fabric dyed in an acidic environment yield exceptional shade of deep red-purple color.

Betaxanthins and betacyanins have different affinities due to the difference in their chemical structures. The latter are less polar than betaxanthins, and this reduced polarity is in favor of fibers. Polar betaxanthins are soluble in water, and thus the equilibrium between the system of dye bath fibers moves toward the less soluble point. Protonation of amino groups of wool and dissociation of the carboxyl groups of both dyes takes place in acidic media that extend their difference of solubility. Protonated dye bears a positive charge in an acidic environment than both dyes (imino groups of these dyes are protonated only in the strongly acidic pH). Wool is attracted by dye on the base of electrical charges [7].

A diffusion phenomenon is a major cause of the different absorption rates of both dyes on the fibers. A lesser flexible and larger betacyanins enter into the nonporous polyamide fiber, much firmer than the smaller and more flexible molecule of betaxanthin, while both dyes penetrate the porous woolen fibers. Furthermore, comparing with polyamide, a woolen chemical variety played an important role which formed the bonds with only a small number of free carboxyl and amino groups terminating long string of these aliphatic chains [8].

Different concentrations and blends of vacuolar pigments produced nearly every color in foliage of plant taxa; for instance, purple and red colors are present. Many other factors affect the flowers, leaves, stems seeds, fruits, and/or roots of plants; nevertheless due to concentration of betalains, color appeared because of the inclusive range of natural environments. The timing of betalain production differs from specie to specie; in the cell vacuole, both betalains and anthocyanins are stored as glycosides, and in dermal, ground, and vascular tissues of vegetative organs, they share similar histological positions. In many cacti, formation of fruit and flower are limited. Anthocyanins remain in senescing organs only for the whole life of the organ but are also present in immature organs. A similar array of red colors was produced by betacyanins and anthocyanins as they are influential antioxidants and produced red color capable of incisive a variability of reactive species.

Photosynthetic pigments (chlorophylls and carotenoids), iridescent structures, and cuticle waxes can also be seen (e.g., iridisomes). Many of these color designs continue enormously in the dark about the adaptive significance excluding this noticeable rainbow of unevenness. Thin band (sometimes<1 mm in width) of orange/red/purple color cells outlines the borders of individual leaves of many plant species. Environmental stress (phosphorus shortage or freezing strain) can create reddening/purpling of the leaf margin in some plants. Constitutive attribute also exists in other species which produces red/purple leaf borders; it continues throughout the lifespan of the leaf and shows its presence through growth of leaf. Since in many taxa, red leaf limits are constitutive traits rather than a strain response, as taxonomists they have used crop species, as a marker to study genetic linkages. Till now, no studies have perceived in case of any biogeographic, phylogenetic, or ecological patterns which resemble with the occurrence or lack of red leaf margins as a constitutive feature. Red pigments may play ecological and physiological roles within the context of the leaf margin. Anthocyanin is accountable for reddening/purpling, and betalains and carotenoids are the basis of leaf margin reddening. The foliage of "Red Tango" sports a reddish-purple center set off by a pink band and frames by silvery green and dark green margins. It is suitable for the baby leaf and mini bunching fresh markets and is suited for year-round production in mild climates. Foliage is oblong, elliptical, and succulent and often has a red-purple spot.

7.6 Stabilization of Betalain Colors

In applications of food colorant, an assessed amount of less than 50 mg betanin/kg could yield the anticipated color. When competing with synthetic dyes, the usage of betalains is greatly reduce due to its poor stability. A list of factors that reduced or enhanced betalain stability is credited to temperature, oxygen, exposure to light, pH, and aqueous and enzymatic activities. Besides this, temperature is the main decisive factor for its decomposition. Microencapsulation technologies can improve the stabilization of betalains.

7.6.1 Factors Affecting the Stability of Betalain

The factors that are influencing the stability of betalain are the temperature, oxygen, light exposure, pH, and aqueous and enzymatic activities.

7.6.2 Heat-Induced Color Changes: The Technical Approach

The extensive heating of red beet or red-purple pitahaya changes its color to red-orange. Decarboxylation and dehydrogenation reactions on heating food were stronger than that of purified solutions. Under optimum conditions, hydrolytic cleavage will result in losing a pigment which is partially reversible. Since C-15 isomerization (i.) and decarboxylation at C-2 (iv.) and C-15 (v.) will not alter the 7ielectron extension, no color changes were induced.

In contrast, carboxyl loss at C-6 reduced the resonance system causing hypsochromic shift of 33 nm resulting in red-orange structures. The most affected color change was due to dehydrogenation at C-14 and C-15, producing neo-derivatives having absorption properties similar to betaxanthins. While neo-formation (14,15-dehydrogenation) along with C-2 and Cn and decarboxylation were the most important reactions for cleansed phyllocactin and hylocerenin, betanin was mainly hydrolyzed into betalamic acid and cyclo-DOPA-glucoside.

Fascinatingly, dehydrogenated and decarboxylated compounds were extra heat stable as compared to their parent compounds explaining why pitahaya juice was more stable than red beet preparations. Further probable structure change that is to be addressed is betacyanin deglycosylation which may be deliberately persuaded by fermentation in the presence of 3-glucosidase. This method has been viable for producing purple tones in red beet preparations. While a distinguished bathochromic shift of 4–6 nm can be observed, further degradation of the aglycones may occur due to oxidative events causing color deterioration resulting in brownish tint.

The worldwide demand for food color was 45,000–50,000 tons in 2009; betalains obtained from red beet fulfill about 10% of the demand. Fabrication of free-flowing betalain powder and current advances in the methods for stabilization enable them to give their greater contribution in natural color market for uses in food, cosmetic industries, and pharmaceuticals.

Addition of preservatives such as oxalic acid may reduce consumer acceptability due to beeturia, a physiological abnormality in 14% of the population to process betalains. On the other hand, stability of betalains can be improved by complexation reaction of betalains with metals, cyclodextrins, etc.; hence, bioavailability of betalain after complexation needs to be explored. Freshly, little studies on inorganic matrices as balanced agents of betalains have been published which include Mg-Al_2O_3, alumina, Zn-AlO, and zeolite for betalain adsorption and exclusion. An analytical tool, like NMR spectroscopy, is used to analyse the adsorption of betalains in c-alumina which can stabilize the pigment through acid-base interaction for more than 20 months. Ceramic particles such as tetraethyl orthosilicate (TEOS) are also used for betalain stability against UV light and temperature to increase its advantages. Carbonyl groups of the betalain molecule formed new carbonyl-oxygen-silicon bonds under acidic conditions. The new compound was comparatively more stable in UV, temperature, and pH, apparently by quenching UV-Vis photons.

FTIR spectroscopy is used for characterizing band, appearing at 530 cm^{-1} possibly representing a C-O-Si group vibration. New structure formation may result in shifting or loss of some of the signals related to carboxylic acid group in the unchanged pigment [9].

Color consistancy is a main concern when they are consumed as food colorant. Stability is affected by several factors such as pH. Betacyanins and betaxanthins have same wavelength in the pH range from 537 to 538 nm of betalain solutions, betacyanin have supreme, whereas in between 475 to 477.497 betaxanthin have extreme. Betanin is stable ranging from 5.5 to 5.8pH in the presence of oxygen.

Vulgaxanthin is highly stable in the range of pH 5.0–6.0 and in purified extracts. It is more constant in liquid at pH 5.7, while optimum pigment constancy on betalain stability is at 440 temperature. Their thermostability is affected by pH solutions and it is reversible. A light-brown color appears by heating betanin solutions, and red color is slowly diminished [10] (Table 7.1).

A nonlinear least-square method used first order kinetics in case of betanin deprivation, under adjustable conditions of time and temperature. A thermal kinetic decomposition model was developed; which allowed to foresee the retention

Table 7.1 Betalain distribution in plant

Factor	Model system	Observation
pH	Betalain absorptivity decreases	Solutions exhibited maximum color stability between pH 3.5 and 7,betacyanins: 537–538 nm, betaxanthins: 475–478 nm pH < 3.5, goes toward lower wavelength, molar absorptivity decreases pH > 7,? Max goes toward longer wavelength, molar
	Betanin solutions with oxygen	Color stability between pH 5.5 and 8
	Red beet solution	Maximum stability at pH 5.5
	Vulgaxanthin solution	Maximum stability between pH 5 and 6
Temperature	Betanin solution	Heating reduces the red color, but cooling may reverse the reaction; degradation follows a first-order reaction
Light	Betanin solution	Rate of betalain degradation increases by 15.6% by daylight exposure at 15 °C; A first-order degradation with pH dependence greater at pH 3 (k = 0.35/day) than at 5 (k = 0.11/day) under fluorescent light. Total pigment destruction by UV radiation or gamma radiation

of betalains. Activation energies (AE) is considered in the range of 17–21 Kcal mol^{-1} for the onward reaction; however, for opposite reaction value falls in the range of 0.6–3.5 pH for betanin decomposition, cyclo-DOPA-5-*O*-glycosideand betalamic acid effected their activation energy. Reverse reaction of the aldehyde group of betalamic acid comprises condensation reaction involving Schiff's base's amine group of cyclo-DOPA-5-O-glycoside where betanin is quickly produced after mixing of both components in solution, which resulted in lower EA and stability of betanin in water-ethanol classical system, where the initial stage of the thermal decomposition of betanin is the nucleophilic reaction on the ring N+ = CH-. Betalain stability is reduced because of its strong nucleophilicity [11]. Its rate of decomposition increases 15.6% at 15 °C when exposed to daylight. When fluorescent light is dropped to betacyanins, decomposition is greater than at pH 5.0 than 3.0.

Alternatively, betacyanins were utmost firm (k = 0.07 $days^{-1}$) at environments. The constancy of betalains and intensity of light range from 2200 to 4400 lux by the absorption of visible light π electrons in the pigment where chromophores are excited to a new high-energy state (π*). Thus molecules will have a greater reactivity or lower activation energy, for instance, EA will be 25 Kcal mol^{-1}in dark and 19.2 in light [12].

Regarding gamma and ultraviolet irradiation effect on betalain consistancy, 17 total pigments are reported which are destructed by the treatments with UV radiation of 120 h or with gamma radiation of 100 krad. The significance of water is manifested in many decomposition reactions, stability of betalains and pigments containing color of a food product are effected by water. Water is responsible for decomposition in diets or representative systems of less moisture. Stability of pigment has an obvious exponential influence, pigment decomposition tracks first-order kinetics, as it declines stability increases from 0.32 to 0.75. Pigment strength rises as water content rises and cannot be itemized only without the water content; color fades, and product becomes dark in the presence of oxygen. Buffered betanin solutions (pH 7) at 15 °C were kept for 6 days in atmosphere of air and nitrogen, upto 15% color decomposition increases because of air environments [13].

First-order model is chased by decomposition kinetics underneath air atmospheric condition but deviates when oxygen is absent. In betanin, deviation obtained from the first-order decomposition kinetics in the lack of oxygen is begun by reaction reversibility. Several methods are described such as removal of gas, stabilizers and antioxidants addition, pH regulator, and treatment of slight health; these attempts help to save from destruction [14].

In spite of their superior antioxidant activity and coloring capacity, betalains do not measure as potential additives in food industry. Their instability prevents long-term storage. The major factor to consider the use of these pigments as colorant in food is stability. Microencapsulation is pronounced as a technique where a bioactive compound is condensed in a biopolymer. One of the objectives of microencapsulation is to increase the shelf life of the bioactive compound, shielding it from unwanted environmental conditions (among others light, moisture, and oxygen) and reducing its reactivity to its outside environment. This method is also used to change liquid solutions to powders, which are easier to handle.

Due to betalain sensitivity, their use as colorants in the food industry is restricted. Grounded on these functions, under low levels of oxygen, humidity, and light, they can be used in foods with minimum heat treatment but with a little shelf life that are packaged and finally marketed in a dehydrated state. They are used as gelatin desserts and in dry mixes, poultry, dairy, confectioneries, and some of the meat products. Problems associated with recapture of pigments and decomposition of betalains during the dispensation operations are resolved by relocating the use of synthetic dyes in some food products [15]. Effective use of cold and different storage atmospheres, practices of handling, systematic enzymatic control, procedure of extraction, purification, to concentrate, and actions for finishing, for instance, freeze, spray, and vacuum drying, helps usage of betalains commercially [16]. Betalain content in red beets are the main focus of recent efforts through discerning breeding. Huge pigment constituents were very vital in the beginning. 130 mg/100 g weight (when fresh) is the average pigment content of beets, but fresh weight of 450–500 mg/100 g is revealed by novel red beet varieties. Besides, as progressive selection is industrialized, this value is increasing [17].

Commercially, food colorants are important in preparation of beet pigments as either powders (that may be produced by freeze drying or spray drying) or fluid concentrates (formed by concentrating juices in vacuum to 60–65% of total solids) [18]. pigment (0.3–1%) is found in them liable on their constituent of yellow pigments, variety of colors is shown by them, beet-like flavor and odor relate to them. The residue of the solids is mostly ash (8–10%), protein (10%), and sugars (75–80%). In laboratory, betalains can be obtained by appointing ultrafiltration, reverse osmosis, and solid-liquid extraction [19].

Efficient recovery of betalains is achieved when it is associated with traditional hydraulic methods. About 80% of beet sap solids are comprised of nitrogenous compounds and fermentable carbohydrates. Fermentation process belonging to *Saccharomyces cerevisiae* and yeast *Candida utilis*) is largely used to remove these materials [20]. A culture of *Aspergillus niger* not only ruined free sugars but also delivered more colorant to betanin. The betacyanin obtained as concentrate from raw juice is concentrated five to seven times than powder obtained from fermented extract (on arid weight basis) [3].

7.7 Extrusion Cooking

Extrusion cooking is the process during which the feed is transported, mixed, and cooked at a high temperature for a small time to obtain extended products, provoking some physical and chemical changes in the feed material for the formation of new products. Variations that are taking place in the feed material through the process are protein denaturation, starch gelatinization, complex formation

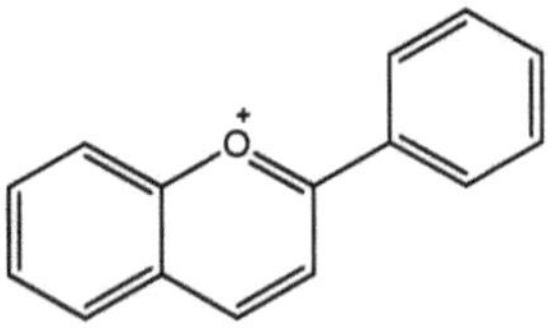

Anthocyanins

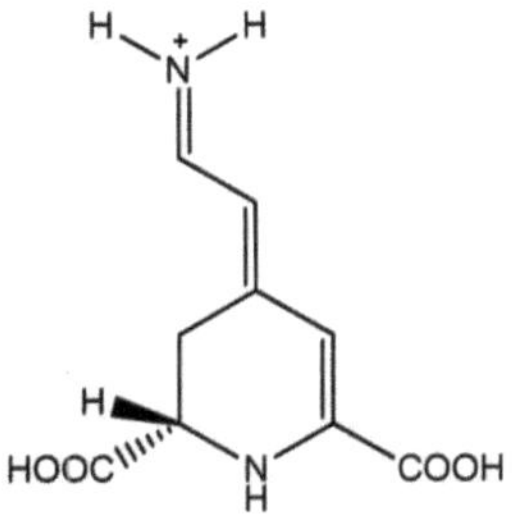

Betacyanin

Fig. 7.1 Anthocyanins and betacyanin pigments

(amylose-lipid), and degradation of heat sensitive components such as vitamins and pigments. These variations are depending upon the process conditions, e.g., types of extruder, moisture content, barrel temperature, feed material, and screw speed among other factors. These operational factors are associated with residence time distribution (*RTD*), a useful tool to evaluate physicochemical changes in feed materials. Moreover, *RTD* is used to scale-up operation. Current extrusion process studies had been engrossed in the development of functional products using legumes, cereals, fruits, fruit extracts, and vegetables. Structural changes of compounds can be seen while using this technology with antioxidant activity such as carotenoids, anthocyanins, and polyphenols, accredited mostly to the high process temperature for pigment extruded cereals; authors investigated the use of fruit powders (blueberry, grape, and raspberry). It was observed that cereals had a good color and anthocyanin and phenolic contents lessened up to tenfold.

The use of encapsulated D-limonene in the extrusion cooking process for increasing the retention of the compound has been reported, to the formation of insertion complexes with starch, gaining products with lesser components degradation. The conduct of the bioactive compounds of encapsulated red cactus pear, for example, betalains, managed by extrusion cooking has not been investigated; therefore, their stability on extruded products should be assessed, because the pigments represent a dye alternative and they are the source of antioxidants in the development of food products. The objective of this study was to evaluate the effect of barrel temperature and screw speed during the extrusion cooking process on bioactive compounds of encapsulated red cactus pear powder (Fig. 7.1).

References

1. Amchova, P., Kotolova, H., & Kucerova, J. R. (2015). Health safety issues of synthetic food colorants. *Regulatory Toxicology and Pharmacology, 73*(2), 914–922.
2. Timberlake, C. F., & Henry, S. N. (1986). Plant pigments as natural food colours. *Endeavour, 10*(1), 31–36.
3. Stintzing, F. C., & Carle, R. (2004). Functional properties of anthocyanins and betalains in plants,food, and in human nutrition. *Trends in Food Science & Technology, 15*, 19–38.
4. Stintzing, F. C., Trichterborn, J., & Carle, R. (2004). Characterisation of anthocyanin–betalain mixtures for food colouring by chromatic and HPLC-DAD-MS analyses. *Food Chemistry, 94*(2), 296–309.
5. Ramírez, E. P., Lima, E., & Guzman, A. (2015). Natural betalains supported on Y-alumina: A wide family of stable pigments. *Dyes and Pigments, 120*, 161–168.
6. Stintzing, F. C., Herbach, K. M., Mosshammer, M. R., Carle, R., Yi, W., Sellappan, S., & Felker, P. (2005). Color, betalain pattern, and antioxidant properties of cactus pear (Opuntia spp.) clones. *Journal of Agricultural and Food Chemistry, 53*(2), 442–451.
7. Simon, P., Drdak, M., & Altamirano, R. C. (1993). Influence of water activity on the stability of betanin in various water/alcohol model systems. *Food Chemistry, 46*, 155–158.
8. Guesmi, A., Ladhari, N., Ben Hamadi, N., Msaddek, M. M., & Sakli, F. (2013). First application of chlorophyll-a as biomordant: Sonicator dyeing of pool with betanin dye. *Journal of Cleaner Production, 39*, 97–104.
9. Khan, M. H. (2015). Stabilization of betalains, a review. *Food Chemistry, 197*, 1280–1285.
10. Wrolstad, R. E., FRANCIS, F., & LAURO, G. (2000). *Natural food colorants. Science and technology* (p. 237). New York: Marcel Dekker Inc..
11. Vargas, F. D., Jiménez, A. R., & López, O. P. (2000). Natural pigments: Carotenoids, anthocyanins, and betalains characteristics, biosynthesis, processing, and stability. *Critical Reviews in Food Science and Nutrition, 40*(3), 173–289.
12. Suh, D. H., Lee, S., Heo, D. Y., Kim, Y. S., Cho, S. K., Lee, S., & Lee, C. H. (2014). Metabolite profiling of red and white pitayas (Hylocereus polyrhizus and Hylocereus undatus) for comparing betalain biosynthesis and antioxidant activity. *Journal of Agricultural and Food Chemistry, 62*(34), 8764–8771.
13. Huang, A. S., & von Elbe, J. H. (1987). Effect of pH on the degradation and regeneration of betanine. *Journal of Food Science, 52*, 1689–1693.
14. Huang, A. S., & von Elbe, J. H. (1985). Kinetics of the degradation and regeneration of betanine. *Journal of Food Science, 50*, 1115–1129.
15. Counsell, J. N., Jeffries, G. S., & Knewstubb, C. J. (1979). Some other natural colors and their applications. *Applied Science London, 2*, 122–151.
16. Lee, Y. N., Wiley, R. C., Sheu, M. J., & Schlimme, D. V. (1982). Purification and concentration of betalains by ultrafiltration and reverse osmosis. *Journal of Food Science, 47*(2), 465–475.
17. Matsuda, N., Tsuchiya, T., Kishitani, S., Tanaka, Y., & Toriyama, K. (1996). Partial male sterility in transgenic tobacco carrying antisense and sense PAL cDNA under the control of a tapetum-specific promoter. *Plant Cell Physiology, 37*(2), 215–222.
18. Wiley, R. C., Lee, Y. N., Saladini, J. J., Wyss, R. C., & Topalian, H. H. (1979). Efficiency studies of a continuous diffusion apparatus for the recovery of betalaines from red table beet. *Journal of Food Science, 44*(1), 208–211.
19. Von Elbe, J. H., & Furia, E. T. (1977). *The betalaines current aspects of food colorants* (pp. 29–39). Boca Raton: CRC Press Inc.
20. Kino-Oka, M., & Tone, S. (1996). Extracellular production of pigment from red beet hairy roots accompanied by oxygen preservation. *Journal of Chemical Engineering of Japan, 29*(3), 488–493.

Chapter 8
Analysis of Betalains

8.1 Introduction

Betalains are frequently present in composite mixtures. They quickly decompose during separation and refining stages that makes separation of higher amounts of pigments difficult. Identification and structural elucidation of betalain molecules comprise of comparison of electrophoretic, chromatographic, and spectroscopic properties and involve direct comparison of these properties with reliable criterions before and after well-organized hydrolysis. The sensitivity of betalains to acid hydrolysis makes them different from anthocyanins, and their color changes with pH changes. In electrophoretic and chromatographic characteristics, both can be distinguished by simple color tests. Since both types of pigments occur exclusively, color tests are usually performed by using crude plant extracts. HPLC is the top-benched method in both quantifiable and semi-preparative works on betalains. The current chapter highlights the history of analyses and the most suitable contemporary methods and techniques employed for characterization (extraction, isolation, and structure elucidation) as well as quantification of these greatly polar nitrogen-containing molecules.

Regardless of equal importance, very little literature exists regarding analyses of betalains as compared to anthocyanin. First successful attempts were carried out in the 1960s for elucidation of structures of betalains, primarily by chemical methods. Betanidin was identified by Wyler and coworkers in 1963 [1], while Piattelli and coworkers characterized indicaxanthin 1 year later in 1964 [2]. Both betalains' subgroups were observed as ammonium analogues of betalamic acid with amino acids or amines as in betaxanthins and with *cyclo*-DOPA as in betacyanins. The sophisticated analytical techniques have now made it easy to identify and characterize the most complicated betanidin conjugates (polyacylated oligoglycosides) like betacyanins from higher plants like *Bougainvillea* bracts [3]. Mass spectrometry (MS) and nuclear magnetic resonance spectroscopy (NMR) are not used to characterize the maximum number of the betalains labeled in the 1960s, and until now these methods

E. Akbar Hussain et al., *Betalains: Biomolecular Aspects*,
https://doi.org/10.1007/978-3-319-95624-4_8

are not practicable to collect exact computable data. Research in MS and NMR has slashed the need of degradation and derivative formation to get structural information of betalains. However, the most popular method of betalain analyses is HPLC.

Many scientists have reviewed the methodologies and techniques which can be used for analyses of betalains formerly [4–7]. The nature and structural elucidations of newly isolated betalains carried out in the past 10 years formerly described in rapports of accepted structures have recently been assembled by Strack and his fellows [8]. The present chapter is an effort to compile the utmost suitable current techniques and methodologies which are used for extraction, isolation, separation, analytical characterization, structural elucidation, and photometric quantification of betalains. Betalains are nitrogenous pigments exhibiting high polarity. A number of proposed betalains structures still complete elucidation. There is ^{13}C record only for one pigment (neobetanin) [9], no heteronuclear NMR analyses for betalains have yet been published, and analyses may be incomplete due to the absence of commercially existing standards (Fig. 8.1 and Table 8.1).

8.2 Extraction

The vegetal substance (e.g., prickly pear fruit/beetroots) must be washed away and sliced. It may then be either mechanically surged to yield betalain comprising fluid, or it may be merged with extraction solvents. Betalains have hydrophilic nature.

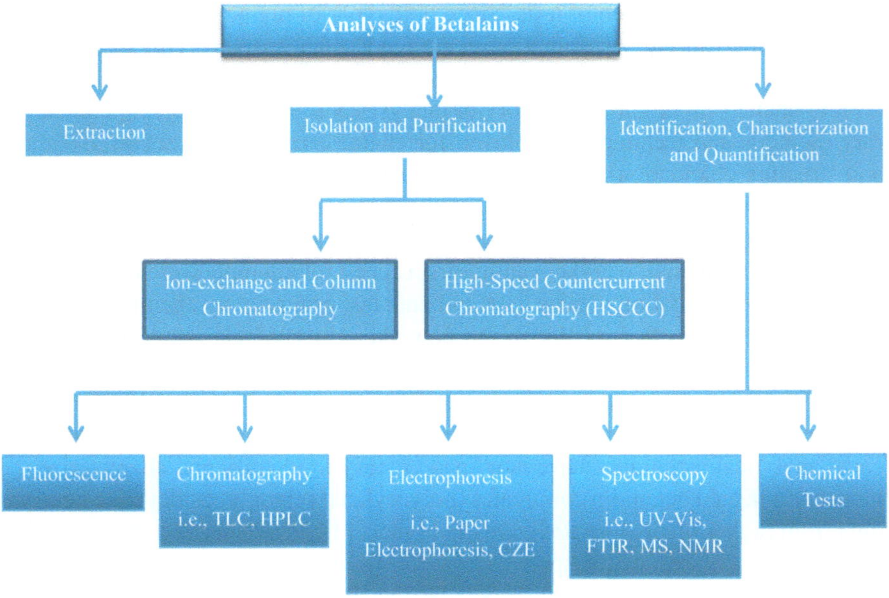

Fig. 8.1 Analysis of betalains

Table 8.1 Methodologies for separation and purification of betalains

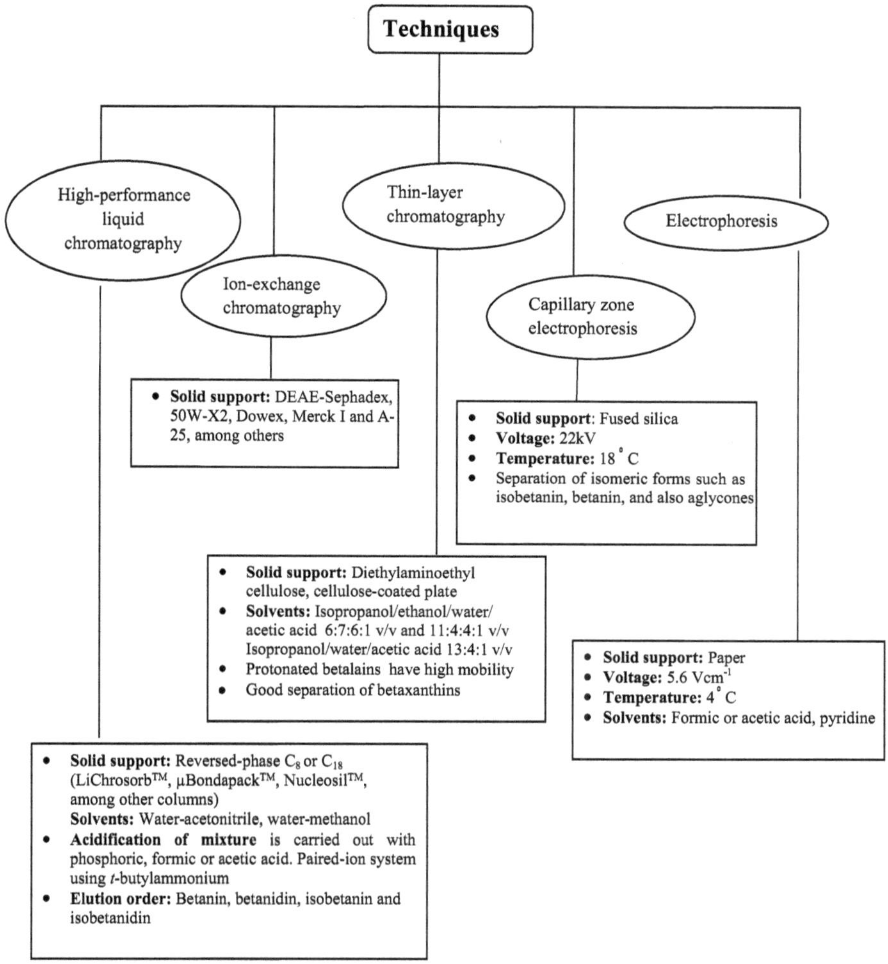

Source: Adapted from Delgado-Vargas et al. [7, 10]

Due to this, all betalain compounds are generally taken out using aqueous solvents like 0.1 M McIlvaine's citric-phosphate buffer solution (pH 5.0) [12] H_2O/0.1% HCl [11], methanol/ethanol (20–50%) and water, or methanol at pH 5 added with 0.25% (w/v) ascorbic acid [13]. A summary of methodologies for extraction of betalains is described in table that is modified from Delgado-Vargas and his coworkers (Fig. 8.2).

Betalains are leached from plant matrix by water, while methanol/ethanol not only ensures complete extraction but also denatures the enzymes present in matrix, thus preventing the degradation reactions of peroxidases and β-glucosidases and ensuring the pattern essentially existing in plant tissue under examination [14–18].

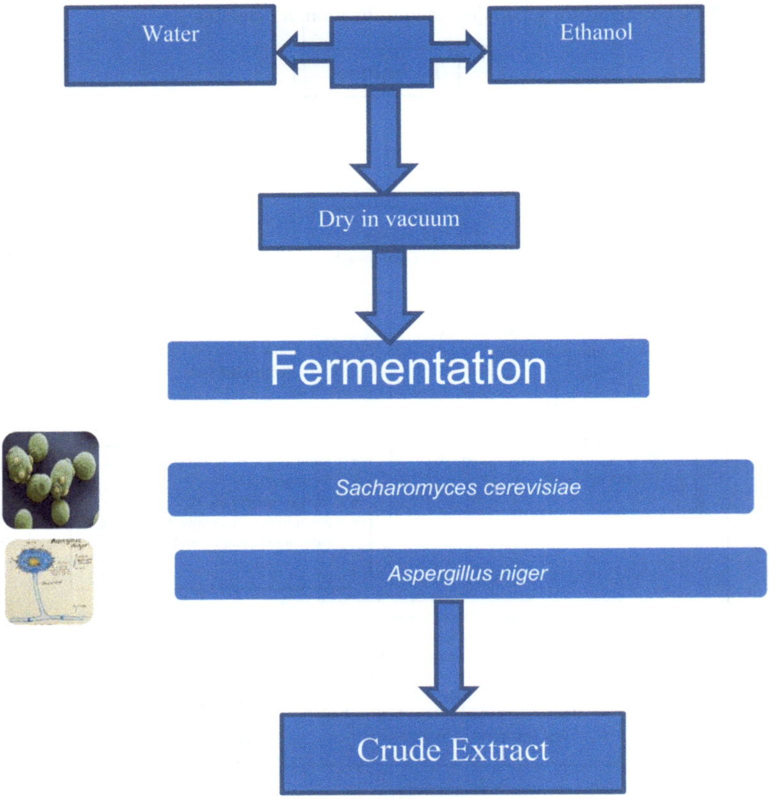

Fig. 8.2 A simplified process for betalain extraction

The suitable ratio suggested is 1 part of plant material and 10 parts of extraction solvent. To prevent oxidation (e.g., browning), sodium ascorbate is added occasionally [19–22].

Since betalains are basic in nature, minor acidification increases purification manyfolds. Minor acidification by HCl or by acidified ethanol (0.4–1%) eagerly precipitates betacyanin [7].

It has been recommended to add 50 mM ascorbic acid in extraction medium which inhibits the possible oxidation by polyphenol oxidases (PPOs) [22]. The reason behind addition of ascorbic acid is that it makes the medium somewhat acidic, which prevents the probable oxidation by PPOs and stabilizes betacyanins. The accumulation of ascorbic acid is obligatory in the presence of great activities of certain betaxanthins (miraxanthin V) and tyrosinase, in the instance of yellow beet hairy root cultures [23]. Then, a readily and nearly whole loss of miraxanthin V happens, supplemented by the presence of artifacts.

However extreme care is needed for acylation of betacyanin with dicarboxylic acids, e.g., malonic acids. Methyl esterification may be carried out by treatment with acidified methanol, thus yielding new molecules which are not part of plant matrix originally [24, 25].

By short heat treatment, many degradative enzymes in extracts having aqueous medium have also been inactivated, but this may however destroy some of the betalain pigments [8].

Maximum betalains were extracted from *Opuntia* fruits by aqueous extraction; however the extract so obtained exhibited highest density and viscosity. Ethanolic extract decreased the viscosity, but betanin quantity was decreased. Betanin yield decreased by 2.6% with 60:40/40:60 water/methanol mixture, while it decreased to 13% with ethanol/ water at 80:20 [26]. The optimum ratio observed was ethanol/ water (60:40) for obtaining *Opuntia* extract with the highest betanin and lowest viscosity [27].

In laboratory, betalains can be extracted by different approaches, such as solid-phase extraction, diffusion extraction, ultrafiltration, and reverse osmosis. All these processes are more efficient than conventional hydraulic techniques for recovering betalains from beet tissues [28]. It is particularly suggested that extraction should be agreed at low temperature and in the dark. The concentration of extract must be done under vacuum at the temperature of bath (less than 30 °C) and should be frozen till evaluation to avoid the decaying process.

8.3 Clarification of Solution

Clarification of primary homogenate is carried out via centrifugation or filtration or sometimes by addition of Celite to the mixture [11]. Re-extraction of filter cube may be carried out until all the pigments have been extracted. Plant fluid (e.g., beet juice) as well as clarified homogenate can be applied openly for high-performance liquid chromatography, spectrophotometry, or electrophoresis; however sometimes there is a need of an extra dilution.

8.4 Isolation Plus Purification

Before qualitative and/or quantitative analyses, purification of betalains is necessary to eliminate all possible interfering compounds present in extract and to enable the fabrication of reference constituents. The intrusive substances may give inaccurate outcomes particularly in spectrophotometric analyses. To evade such improper results, other procedures (mostly electrophoretic/chromatic) are used to isolate and purify betalains from various sources. These processes carry out distillation as well as isolation of pigments, thus letting the quantitation of separable and total betalains altogether.

Since betaxanthins and betacyanins share similar electrophoretic/chromatographic characteristics, therefore parallel methods are applied for their separation and purification. However for isolation of betaxanthins, polyamide chromatography is of little value. Betacyanins involving aliphatic acids consisting of ester linkages have been effectively purified by precipitation through controlled acidification with mineral acids like hydrochloric acid (HCl) [29]. The processing of extracts may be done via ion-exchange chromatography for the elimination of free sugars and organic acids or by precipitation of pectic constituents [25].

Different molar extinction coefficients were reported because of different solvent system purities and water content pigments. Among which, for betalain 60,000 L/mol*cm is more reliable. The following formulas are used to calculate 48,000 L/mol*cm for betaxanthin and 56,600 L/mol*cm for amaranthin contents:

$$\text{mg/g content of betalain (fresh weight)} = \frac{A * MW * V_f * DF}{\varepsilon^* L^* W}$$

$$\text{mg/g content of betalain (dry weight)} = \frac{A * MW * V_d * DF}{\varepsilon^* L^* W_d}$$

$$\text{mg/L content of betalain} = \frac{A * DF * MV * 1000}{\varepsilon^* 1}$$

where the term "MW" represents the molecular weight of vulaxanthine I (339 g/mol), amaranthine (726 g/mol), indicaxanthin (308 g/mol), and betanin (550 g/mol), respectively. "A" symbolizes absorption, corrected by absorption at 600 or 650 nm, "W" is the fresh (W_f) or dry weight (W_d) of the plant material, "V" is the total extract volume, and molar absorptivity "L" equals path length where "DF" is the dilution fector.

8.5 Ion-Exchange and Column Chromatography

Piattelli and Minale developed the first standard method for betalains analyses [30]. In ion-exchange resin, aqueous extract of plant was added due to which nonionic adsorption of betalains on resin takes place. Then, 0.1% aqueous HCl is used to wash the resin, and elution of pigments is carried out with water followed by separation of individual constituents on a polyamide column.

In fractionation, such as separation, the most commonly used adsorbents are ion-exchange resins than gel filtration. The purification and separation procedure is completed by ion chromatography on Sephadex G-25 (gel filtration) or Dowex 50 W (cation-exchange resins) followed by subsequent adsorption chromatography using Polyclar AT (polyvinylpyrrolidone) and/or polyamide [31] employing developing solvents such as elevated methanol concentrations in citric acid [30]. From betalains, the acid is then removed by resin action. By applying this method, betacyanins from flowers of *Lampranthus sociorum*, cell cultures of *Chenopodium rubrum* [32], and flowers of *Gomphrena globosa* [33] have been separated (Table 8.2).

Table 8.2 Selected application data for (semi-)preparative reversed-phase HPLC of betalains

Application	Betanin[a]	Betanin and its glucosides	Humilixanthin	Betanin[b], gomphrenins
Column[c]	μBondapak/Porasil B (35–75 μm; 610 × 7.8 mm i.d.)	OPTI-UP C12 (65 × 25mm i.d.)	Multisorb C_{18} (10 μm; 250 × 20mm i.d.)	Silica C-18 (10 μm; 300 × 40mm i.d.)
Solvent[d]	Potassium phosphate/ methanol/acetic acid 0.05 M (81.2:17.8:1)	Aqueous acetic acid (1:100 mM)	Gradient with formic acid/ acetonitrile (0.4:50%)	Gradient with formic acid/ methanol (1:80%)
Flow rate	8 ml/min	4.5 ml/min	5 ml/min	20 ml/min
References	Schwartz and von Elbe [51]	Trezzini and Zryd ([77, 78])	Strack et al. [49]	Heuer et al. [33]

[a]In addition, isolation of betalamic acid and cyclo-DOPA-5-*O*-glucoside was achieved; these were obtained after heating betanin. (Schwartz and von Elbe [79])
[b]See Fig. 11.1
[c]Trade names of commercially available packing material
[d]*i.d.* internal diameter

Before gel filtration chromatography, the crude extract pH value should be adjusted with HCL to 3.0 [34]. A new procedure comprises of Sephadex G-25 [35], followed by Sephadex LH-20 chromatography prior to preparative high-performance liquid chromatography has been proved very beneficial [36]. Similarly, for the semi-purification and concentration of samples of betalains as well as for isolation of betaxanthins from betacyanins, C-18 sorbent using solid-phase extraction is also utilized employing pH-based retention features [25].

Alternatively, stirring of betalain extract is done with the ion-exchange resin (for instance, Merck I, DEAE-Sephadex A25, Dowex 50 W-X2, etc.), which adsorbs betalains due to nonionic interactions. Afterward, resin is washed with 0.1% aqueous HCl (v/v), and elution of betalains is done by using water as eluent, and then absolute separation is carried out by using a polyamide, polyvinylpyrrolidone, Polyclar AT, or Sephadex G-15- and G-25-type chromatographic columns. Consecutive chromatography using a series of Sephadex ion exchangers is the best purification and isolation technique for betalains [35, 37, 38].

The following points are of cardinal importance when working with polyamide resins:

1. Retention of 5-glycosides precedes that of 6-glycosides.
2. Increased glycosyl substitution decreases retention of betacyanins.
3. Retention of iso-derivatives is somewhat longer than the corresponding C-15 epimers.
4. Retention of pigments increases with acylation and aromaticity.

The anion–/cation-exchange resins require high salt or acid concentrations for fractionation and purification. Addition of salt can substitute sugar against buffer or NaCl in resulting elutes requiring an additional desalting step. Similarly addition of acid can degrade the betalain molecules especially during the evaporation process.

These demerits coupled with long analyses errors, traditional CC, high chemical consumption, and low selectivity have currently been typically swapped by high-performance liquid chromatography.

8.6 High-Speed Countercurrent Chromatography (HSCCC)

In a recent paper, HSCCC has been used to isolate betalain pigments for preparative purpose from the extracts of red beet. A highly polar solvent system comprising of ethanol/acetonitrile/saturated solution of ammonium/water (1.0:0.5:1.2:1.0), due to the hydrophilic character of betalains, was engaged for pigment isolation in a marketable juice of red beet and concentrated deprived of additional sample work-up. Betanin and its C^{15} epimer (isobetanin) are exhibited by the main peak [39].

8.7 Fermentation

Fermentation as well as caramelization occurs due to free sugars present in beet extract during food processing at high temperature, thereby increasing the degradation of betalains. These sugar-rich extracts decrease tinctorial power; therefore separation of sugars from beet extract is necessary [7]. Therefore beet juice is yeast fermented to remove fermentable carbohydrates and nitrogenous compounds which make up 80% of juice [40]. This fermentation decreases sugar contents and increases the betalain contents.

Polyethylene glycol 6000 (PEG) along with ammonium sulfate ($(NH_4)_2SO_4$) is the most appropriate system for purification as sugars (80–90%) settle down to lower phase, while betalains (70–75%) are partitioned into upper phase. Later PEG is removed from betalains by chloroform/water extraction. The sugars may also be removed by aqueous biphasic extraction [41, 42]. When intrusive substances such as fresh juices, extracts, and refined samples are not present, the quantification is completed by means of spectrophotometry whereby absorbance at evident maximum is expressed as concentration by using suitable absorptivities [43–45].

8.8 Identification, Characterization, and Quantification

By comparing the spectroscopic, chromatographic, and electrophoretic data of betalain with the data of standards, structure elucidation of known betalains is easy to carry out [25, 46]. After proper hydrolysis by standard testimony of amino acids, sugars, and acyl moieties, additional structural properties can be accomplished [5, 25]. By the reaction of amino compounds with betalamic acid, the standards of betaxanthin may be prepared [25, 47].

8.9 Fluorescence

Fluorescence of betalains has been explored by very few researchers, but recently, a structural associations of fluorescence related with betaxanthins study was done, but not with betacyanins. In case the of 1,7-diazaheptamethin, the fluorescence may be due to electric resonance, but the fluorescence was gone in the case of betacyanins due to resonance extension toward the ring of indole moiety. In addition, the presence of electron withdrawing group such as carboxyl group tends to enhance the intensity of fluorescence whereas electron donating groups such as hydroxyl and aromatic ring tend to reduce it [48]. Methylation with diazomethane, yielding derivatives (λmax 340–360 nm) confirmed the existence of the 1,7-diazaheptamethin system in all betalains that showed bathochromic shifts (80–100 nm) in acidic solution [46].

8.10 Thin-Layer Chromatography (TLC)

Thin-layer chromatography (TLC) is not extensively used for betalain analyses due to low R_f values. However, a preparative TLC system was developed by Bilyk in a 0.5 mm cellulose-coated plate on which two distinct mobile phases were used: first solvent mixture having isopropanol-ethanol-water-acetic acid in a ratio of (6:7:6:1) and for the second mixture in a (11:4:4:1) ratio. The protonation of carboxyl group of betacyanin and increment in mobility of betalain on the TLC plate take place by addition of acid in developing solvent. Acid anion provided the electrically unbiased organization due to its collaboration with quaternary nitrogen. Betaxanthins also exhibit the same effect.

From the bracts of *Bougainvillea glabra*, separation of a composite mixture of pigments (betacyanins) has been done by TLC on cellulose, using solvent (ethyl acetate-formic acid-water) in ratio of (33:7:10) [3]. Betaxanthins have also been separated on diethylaminoethyl (DEAE) cellulose [49]. There is no need of indicator for the visualization of separated pigments due to their natural colors. Due to high polarity of betalains, TLC has been reported as a separation technique though the separation is made advanced via addition of acids [50].

8.11 High-Performance Liquid Chromatography (HPLC)

When products of degradation and the other intrusive materials are present in mixtures, HPLC is the best technique for the chromatographic isolation, quantification, and the tentative identification of betalains [51]. Vicent and Scholz applied it for the first time by gradient run in a C_{18} column using mobile phase of tetrabutylammonium in the paired ion system.

Red and yellow pigments were monitored by using binary wavelength sets at 538 and 476 nm. 476 nm is the wavelength of choice for betalains' comprehensive screening due to absorbance of red betacyanins at a certain degree at 476 nm. Yellow betaxanthins are eluted earlier than betacyanins because they are more polar due to their chemical structure. For quantitative analyses, calibration of HPLC system was done with a purified standard of betanin that resulted in excellent linearity over a practical work range of 0.005–0.03% betacyanin (w/v).

Stereochemistry and overall polarity are two principal factors for separation of betalains. Betacyanins usually elute later than betaxanthins. Number and nature of involved sugars as well as sugar acylation and site of attachment are of special importance in betacyanins. Consequently, betanin (betanidin-5-O-glucoside) shows a lower retention time as compared to gomphrenin I (betanidin-6-O-glucoside) [33].

Betacyanins are usually found as C-15 epimers having greater number of betanidin derivatives in contrast to corresponding stereoisomers [52]. β-Glucosidase hydrolysis or slight acid hydrolysis [53] produces only betanidin, while strong acid hydrolysis produces a mixture of both aglycones, betanin, and glucose [54]. S and R C-15 epimers of betacyanins can be separated easily as S-form elutes first. By sensitive detection at lower wavelength of 465 nm and the absence of typical second R-isomer peak, ease of distinction of neobetacyanins, for example, neobetanin, present in fruit of *Opuntia ficus-indica* is made [49]. Betaxanthins are usually detected between 475 and 480 nm, while betacyanins are detected between 535 and 550 nm. Elution time is increased via acylation of hydroxycinnamic acid in the same order in which the polarity of ester moiety tends to decrease. Acyl number of betacyanin may be determined by adjusting to wavelength to favor the recognition of the acyl moiety. Peak ratios of absorbance are about 2:1 at 535–540:315–325, respectively, for monoacylation.

C-11 epimers of betaxanthin (R-and S-indicaxanthin) were separated for the first time by Terradas and Wyler [55] regarding betaxanthin epimers. Contrary to the isomeric betacyanins, S-indicaxanthin manifests a longer retention time than R-indicaxanthin. Till now, observation of 11R-indicaxanthin is not made in the universe.

Acylation protects sugar moiety from enzymatic cleavage. Only isobetanidin stereoisomers are produced from acid hydrolysis of isobetanidin derivatives [30]. Therefore by using a mixture of both molecules, subsequent analyses can be easily accepted to conclude whether two unknown structures of betacyanins are C-15 epimers or not [52]. Alkaline hydrogen peroxide oxidation can release basic glycosyl molecules besides acid and enzymatic treatments [56]. HPLC or paper chromatography can be used to identify the cleaved sugar group, while aglycone moiety and the products resulted from its degradation can be recognized by conservative approaches, e.g., by HPLC or paper chromatography, in which it is compared with authentic standards.

Treatment with excess diazomethane can be used to identify the position of glycosyl attachment in which O-methyl neobetacyanidin trimethylester glycoside is formed aglycone moiety. In O-methyl neobetacyanidin trimethylester glycoside, originally free phenolic group of aglycone becomes methylated and thus is easily identified [52]. Non-acylated pigments possess weak absorption, while acylated ones usually show a second absorption maximum in ultraviolet region of 260–320 nm [46, 52, 57]. The visible maximum absorption ratio at ultraviolet maximum assesses acyl residues amount in betalains [52]. The type and position of sulfate, sugar, or acyl groups attached to glucosyls are usually estimated by nuclear magnetic resonance spectroscopy.

Esterification of sugar such as glucose with ferulic acid at sixth carbon causes extra absorbance bands of gomphrenin III (324 nm) and lampranthin II (322 nm) that were observed in spectra of gomphrenin I and III and betanin and lampranthin II. The visible maxima were shifted toward higher wavelengths (8 nm for gomphrenin III and 5 nm for lampranthin II) relative to non-acylated molecules. This is explained on the basis of interaction between the acyl group and betanidin nucleus [36].

Silica (SiO_2) is the commonly used column support, whose OH are derivatized with octyl or octadecyl silyl groups (particle size = 3–10 μm; C8 or C18 reverse phase) [5]. Huang and von Elbe separated and purified amaranthin and betanin from amaranth leaves using a grouping of paired ion chromatography and gel filtration [34]. Regardless of all the effective outcomes, a number of betalains' HPLC investigations are accomplished in the absence of ion-pair reagents. The mobile phase is usually an acid or buffer such as acetic acid or phosphate buffer which ensures a low pH value such as 3.0 and an organic modifier (e.g., acetonitrile, methanol), while RP-C18 [51] or RP-C8 [58] columns are almost exclusively used as the stationary phase [59]. Sensitivity of HPLC methods depends on pH value as best results are obtained with pH 4 [60, 61].

In HPLC, pure crystalline pigments are eluted in the following order: betanin, betanidin, isobetanin, and isobetanidin. This suggestion depends on the glycosides acid hydrolysis in order to produce aglycones and isomerization of betanin to isobetanin.

By using a C_{18} reversed-phase column and ion-pairing using methanol-water mobile phase, a fermented red beetroot extract was analyzed more recently [40]. The elution order was betanin, isobetanin, betanidin, isobetanidin, and prebetanin for the betacyanins and vulgaxanthin I followed by vulgaxanthin II for the betaxanthins. Another good application of characterization of betaxanthin was carried out by [47]. A series of betaxanthins were yielded by the conjugation of betalamic acid with amino acids both proteinaceous and non-proteinaceous. Retention characteristics were defined by them for 15 naturally occurring betalains such as vulgaxanthin I, miraxanthin II, and portulaxanthin I, among others. For the unknown pigments, such products could be utilized as HPLC standards.

UV-Vis-coupled HPLC detects peaks 535–540 nm for betacyanins and 470 nm for betaxanthin. Diode array detector monitors wavelengths at 406 nm for betalamic acid, 280 nm for colorless phenolics, 470 nm for betaxanthin, and 536 nm for betacyanins. Complete separation is usually achieved via gradient elution.

For betalain analyses, photodiode array detector is commonly used [33]. Scanning of the sample is done very rapidly after every few milliseconds producing ultraviolet-visible spectral data and manipulating the maxima of absorbance. By calculating the absorbance ratio at two different wavelengths or by multiple absorbance ratios determined at whole range of wavelengths relative to the wavelength of maximum absorbance providing various spectra of potential impurity and analyte, the purity of peak can be checked. It has led to easy detection of betacyanins acylated with hydroxycinnamic acids.

HPLC has been useful tool for measurement of enzyme activities during betalain biosynthesis. Easy product identification and quantification have been observed in studies on betanidin glucosyl transferase [62], betacyanin acylation [63, 64], and extra diol cleavage of L-DOPA (L-dihydroxyphenylalanine) during the development of muscaflavin and betalamic acid via 2,3-secodopa and 4,5-secodopa, respectively [55].

Generally, the HPLC separation of betalains can be accomplished within 20 min. Up to now, with the aid of HPLC, little investigated betalains have been detected, and useful evidence about the degradation and stability of a number of betaxanthins and betacyanins has been collected [51, 65]. HPLC has become the selected method for quantifiable analyses of individual and total betalains due to its quickness, high-resolving power, good sensitivity, and not always essential primary purification point [66].

The advantages of this technique like speed, high separation capacity, ease of quantification, and good reproducibility render it a chief analytical method for betalains in the future. Due to the evolution of advanced micro- and many other columns and stationary phases such as microbore material, HPLC separations can be improved and through which also lower the required consumption of solvents and the sample volume. The online characterization of eluted pigments has been improved by the employment of more sophisticated detectors such as electrochemical and diode array detectors. HPLC-ESI-MS is the most wanted methods for unknown betalain determination in complex mixtures. Moreover, it is also applicable for quantification as well as clear characterization of betalains in prickly pear [49], beetroot [67], and other well-studied products. Despite its merits, HPLC has its limitations. Many compounds present in trace/minor amounts can go undetected although they contribute to overall pigment contents. Similarly many standards are required if sample contains more than one major betalain molecule.

8.12 Electrophoresis

Betaxanthins and betacyanins possess quaternary nitrogen atom with weak positive charge and three COOH groups. The nitrogen atom in combination with two-position COOH group imparts amphoteric properties to both compounds; therefore they ensue as negatively charged ions at pH > 2 and travel in electric field. Indicaxanthin and betanin have been associated with electrophoretic mobility of betaxanthins and betacyanins, respectively.

Electrophoretic separation of betalains is mostly performed by using various paper strips (Beckman No. 319328, Whatman No.1, S&S 4043B) in paper electrophoresis. Various factors like pH, temperature, electroendosmotic flow, buffer concentration, and composition influence separation to a larger extent. Sharper bands are obtained at low temperature (4C), and current gradient between 0.5 and 2 mA/cm and 5.6–13 voltcm^{-1} voltage gradient set [68, 69] obtained best bands by using 0.1 M phosphate buffer at 5.5 pH. Vulgaxanthin I and prebetanin are better separated by including pyridine as base in buffer solutions [70]. The immediate drying of papers is carried out at 30 °C or 80–90 °C in vacuum oven [68, 69], while $CaCl_2$ is added sometimes [71]. The different color bands are marked in daylight, and compounds are identified by relative migration length or by coelectrophoresis with standard betalain pigments. Densitometer is used to scan the band areas of compounds and calculation of values is done as betanins for quantitative determination. During measurement of betacyanins of red beet, standard deviation of 0.4 mg within range of 2–9 mg betanin was observed [71]. Golden and yellow beet betalains were effectively distinguished into eight compounds, the utmost significant of which are vulgaxanthin I and II [68]. Although paper electrophoresis is a simple, speedy, and commonly used technique in betalain analyses, it fails where complete separation and identification of betalains are needed. Paper electrophoresis was supplemented by subsequent paper chromatography for complete separation of betaxanthins. Betacyanins can be completely separated by column chromatography (polyamide) followed by electrophoresis [69].

Paper electrophoresis on cellulose using formic or acetic acid and pyridine solvents is the common and most reliable method for the detection of betacyanin, because firstly they travel as immobile zwitterions at 2 pH, then as monoanions at 2–3.5 pH, and at the end as *bis*-anions having 3.5–7.0 pH. The mobility may be associated to indicaxanthin in the case of betacyanins and betaxanthins are associated with betanin mobility. At 5.6 Vcm^{-1} voltage, gradient electrophoresis can be accomplished using pyridine-citric acid as solvent and at a temperature of 4 °C.

Reverse electroosmotic flow capillary electrophoresis is also a right practice for betalain analyses [72]. On a silica capillary tube, chief betacyanin compounds of beetroot extract were separated by using citric acid-dipotassium phosphate buffer containing dodecyltrimethylammonium bromide (6 mM DTAB) and Ca ions (4 mM) at room temperature and voltage of −20 kV.

To analyze betalains particularly from the source like *Beta vulgaris*, capillary zone electrophoresis (CZE) with multiwavelength has been used. It is beneficial with a fused-silica capillary at constant voltage of −22 kV and 15 °C temperature that can easily separate isobetanin, betanin, and their analogous aglycones. For separation and exposition of betalains, CZE has been efficiently used; although it takes longtime evaluation, if only two major pigments are focused to be separated, leaving the aglycones unseparated, time can be reduced. CZE with an acceptable resolution can also be used for the quantification of betaxanthins. Remarkably, it has been observed that betalain quantification results obtained by CZE are in close agreement with that of HPLC results.

8.13 UV-Visible and FTIR Spectroscopy

The analyses of betalains have been centered principally on UV-visible spectroscopy as that of the other colored compounds. Due to the conjugated dienes of 1,7-diazaheptamethin substructure and phenolic character of betalamic acid, all betalain pigments show absorption maxima in visible and ultraviolet regions, respectively [30, 48].

Indeed, the chief betacyanin of beetroots, betanin, showed maximum absorbance at 535–540 nm with an absorptivity value of E1cm 1120 in its absorption spectrum [30]. Similarly, yellow-colored vulgaxanthin exhibited absorption maximum at 476–478 nm with E1cm = 750 [73].

Nilson developed a spectrophotometric method for synchronized determination of yellow (vulgaxanthin I) and red (betanin) in beetroot juice [69] without preliminary isolation of these compounds. Since a two-compound mixture shows absorption spectrum that is cumulative of each individual compound spectrum, therefore absorption of one compound should be subtracted at spectrum of the second. In this specific case, vulgaxanthin I exhibits specific absorbance at 476–478 nm only. It was reported that at 535–540 nm that is characteristic wavelength for betanin, vulgaxanthin I did not show any absorbance, whereas betanin can absorb at maximum absorbance of vulgaxanthin I, i.e., 476–478 nm, revealing that absorption at 535–540 nm is characteristic of betanin. By calculating the ratio of absorbance at 538 nm to that at 477 nm as function of concentration (dilution), a measurable concentration of vulgaxanthin I was determined. Direct sample determination is mainly due to the sample absorbance at 538 nm. Results are expressed as total betaxanthin and total betacyanin since other betalain constituents present in minor amount also contribute to measured absorbance.

Saguy et al. also developed an accurate and rapid method for determination of all major beet pigments (vulgaxanthin I, betalamic acid, betanin) [74]. The method is superior, and computer-aided quantification is grounded on spectrum nonlinear curve fitting with an expected individual compounds' function.

The method has ended time-consuming and laborious separations leading to continual monitoring of temperature- and time-related procedures. The outcomes obtained by this process are in outstanding conformity with natural betalain contents as compared to Nilsson method. The yellow betaxanthins absorb at 480 λmax, while red violet betacyanins absorb at 540 nm.

When betacyanins are acylated with the aromatic acids, a second absorption maximum is observed at 300–330 nm range, whereas non-acylated betacyanins show weak absorption. With the structure and solvent and even by changing concentration, the molar absorptivities (e) of betalain pigments fluctuate significantly. The e-values like 48,000 and 61,000 may be used for quantification purposes for the isolated and purified betaxanthins and betacyanins, respectively, noted at their visible absorption bands [75]. Absorption bands in Fourier transform infrared (FTIR) spectra may be used for the characterization of typical functional groups of the betalains [76].

Due to cross-absorption of analyte and associated compounds, the spectrophotometric determination may lead to underestimation and/or overestimation particularly in the samples to which heat is provided, while IR spectra of betalains provide very little information. Although spectrophotometric methods are simple, speedy, and cost-effective, these are insufficiently selective.

8.14 Analyses of Betanin Sources

Fresh juice of beetroot, commercial betanin diluted with dextrinase, food-grade beetroot (i.e., bee powder), and commercial lyophilized (freeze-dried) beetroot are normally used for analysis. These are some commercially applied betalainic sources in North America and Europe for coloring purposes.

Every spectrum was elaborated into a Gaussian line having 536 nm center, supreme height fixed at 478 nm of split-Gaussian line was seen and a non-constrained split-Voigt line, correspondingly, to the betaxanthins (Bx, kmax = 478 nm, e480 = 4.8–104 L mol^{-1} cm^{-1}) isobetanin-betanin mixture (Bns, kmax = 536 nm, e535 = 6.5–104 L mol^{-1} cm^{-1}) and other components (kmax <300 nm, counting browning materials exhibiting 600 nm absorbance). The spectral fitting of >600 nm 350–420 nm regions is enhanced when related to earlier outcomes. Deconvoluted band analyses designate that the highest relative amount of Bns is present in sample A, i.e., Bns/Bx molar ratios: 1.4 (sample C), 1.1 (sample B), and 2.3 (sample A).

UV-Vis coupled with RP-HPLC analyses (254 and 536 nm) and MS (ESI+, m/z 200–600) were carried out in samples. Deconvoluted bands relate to Bx (yellow) and Bns (magenta), and fitted spectra are displayed by continuous red line, r2 = 0.9992 (sample C), r2 = 0.9994 (sample B), and r2 = 0.9991 (sample A).

In mg/100 g of raw material, betanin contents were compared via UV-Vis spectrophotometry, and RP-HPLC/UV-Vis is given. Species absorbing at 536 nm the concentration in RP-HPLC elution (LCBns+) was found by assuming e = 6.5–104 L mol^{-1} cm^{-1}, to compare it with UV-Vis spectrophotometry (VBns+). The concentrations of isobetanin (LCiBn, 6.4 ± 0.2 min) and betanin (LCBn, tR = 6.1 ± 0.2 min) were found by calibration curve in figure below.

By spectrophotometric and chromatographic means, the inconsistencies in the quantification of betalains can reach 15%. Direct absorption measurement at 536 nm caused the betanin concentration in overestimates of 4% for sample C, 25% for sample B, and 8% for sample A (lyophilized beetroot). This was due to the Bn breakdown into oxidized (i.e., neobetalains) and decarboxylated (at C2, C15, and C17) analogues at 536 nm absorbance during the process of lyophilization in addition to the huge betaxanthin amount captivating at 480 nm in sample Bs. Discrepancy for sample B is around 9%.

Though sample B is a commercial product, sample browning occurs via sample A lyophilization directly after the extraction of juice (starting pH 6), perhaps because of the increasing polyphenol oxidase enzyme (PPO) concentration via freeze-drying. Furthermore, during concentration due to oxygen, metals, lack of pH control, or light mostly leads to breakdown of betanin.

Results showed that by spectrophotometric methods, the mixtures of isobetanin/betanin can only be clearly calculated in case when the quantity of other materials showing 400–480 nm absorbance and at 536 nm is reduced [42].

8.15 Amino Acid Analyses

Betalain analyses also occur through amino acid analyses. Their hydrolysis occurs with 0.6 ammonia or 1 N aqueous HCl to get free amino acids and betalamic acid. Synthesis of betaxanthin takes place by betanin reaction with an excess of amino acids in ammonia alkaline solution. A betaxanthin increment is tested extreme at 475 nm absorption or decrements of betanin maximum at 540 nm. By mixing betanin with an excess (10 M) of glutamic acid in 0.6 ammonia solution, vulgaxanthin is gained; to synthesize betaxanthins such as miraxanthin and indicaxanthin, this base exchange method can be used.

Nonnatural betaxanthins could be attained with lysine, threonine, and serine. Methylneobetaninidin dimethyl ester shows absorbance at 403 nm which can be shifted by acid addition from 513 to 375 nm [7].

8.16 Mass Spectrometry

MS technique was first time used on flowers of *Gomphrena globosa* for quantitative analysis of betacyanins present in it, and it is a suitable technique for thermolabile compounds like betalains. Electrospray ionization tandem-mass spectrometry (ESI-MS-MS) is the best method of choice for identification and characterization of structurally related betalain pigments [33]. Scans of parent ion by selecting the m/z 389 fragment ion, i.e., [betanidin + H^+], are normally carried out for a thorough identification of isobetanidin derivatives' protonated molecular ions $[M + H]^+$ [33]. For crude extract analyses, the parent ion scan method is largely well-suited giving a better signal-to-noise ratio after HPLC separation coupled with tandem mass spectrometry detectors. A micro-HPLC system having time-of-flight (TOF) mass detection and electrospray ionization method is used to achieve the accurate masses of betalain pigments from the different *Celosia* species [21]. Ultrahigh mass resolution is associated with Fourier transform ion cyclotron resonance mass spectrometry (FT-ICR-MS). Hence it is very advantageous for betalain structure elucidation.

It might be an easy chore first to enhance pigment separation trialed by mass spectrometric measurement. Mass spectrometry is also used to determine structure.

8.17 NMR Spectroscopy

Nuclear magnetic resonance spectroscopy is a useful technique; as for betaxanthins and betacyanins, H^1-NMR data have been reported, and for C14–C15 saturated betaxanthin and betacyanins, ^{13}C data have only newly become accessible. Highly acidic settings were detrimental for betacyanins as noted by previous reports. For 14,15-dehydrobetanin (neobetanin), the first ^{13}C NMR data were obtained, showing an experimental stability toward low pH values.

So far little literature exists on ^{13}C NMR of betalains and requires extensive analysis on more sensitive spectrometers to have better understanding of its structure in the future. In all NMR studies, it is recommended to acidify solution to get stable zwitterion of betalains in order to obtain clear spectrum. ^1H NMR data on betalain pigments have been documented in deuterated methanol, dimethyl sulfoxide, and acetic acid, the bits of hydrochloric acid. Conversely, solvents comprising acids, chiefly the mineral acids like deuterated acid, may hydrolyze the bonds, in sugars and aliphatic acyl moieties. It may also degrade products, and exchange of deuterium can occur for anticipated NH and OH groups as well as for hydrogen bonded with carbon. Therefore, in betacyanin system, H-12/H-18 undergoes exchange, while on extended exposure, H-7/H-15 may partially exchange. Furthermore, some carbon-linked protons are difficult to detect because they are exchanged with deuterium present in the solvent. Shifts of H^1 NMR on betalains are tough to allocate on the basis of only ^1H NMR 1D spectrum. More recently, homonuclear 2D techniques such as ^1H-^1H DQF-COSY and ^1H-^1H TOCSY have afforded valuable information for the assignments. The nature of aglycone moiety and number and nature of sugar molecules and acyl constituents can be determined by COSY (1D and 2D shift-correlated) ^IH NMR spectroscopy.

Strack et al. considered a convenient compilation of ^1H NMR data 5. ^1H NMR data on the infrequent 2-descarboxybetanidin and betacyanidin are shown in Fig. 10.17. Aglycones of betacyanin occur either as 2-descarboxybetanidin, betanidin, or isobetanidin (Figs. 10.16 and 10.17). A 2S and 15S betacyanin seems to have H-14, H-12, and H-11 A at 0.02–0.04 ppm that is slightly higher fields than analogous 2S, 15R isomer. A two-proton triplet around 4.4 ppm in 2-descarboxybetanidin supplanted the double doublet of H-2 around 5.6 ppm of betanidin and isobetanidin [21]. At the partial C-12/C-13 double bond, betacyanins are recognized to exist as E/Z isomers. Compared to those of 12 E-isomer, H-11 and H-12 signals of 12 Z-isomer are usually recorded at higher and lower fields. The position of glycosyl linkage to either fifth or sixth carbon of betacyanidins should be discovered clearly by HMBC spectra in which a long-range coupling between the aglycone C-5 or C-6 and anomeric sugar proton is observed. A low chemical shift of difference ~ 0.1 ppm between H-4 and H-7 is, however, distinctive for substitution of glycosyl at the hydroxyl group of betanidin at carbon 5, contrasting to ~ 0.8 ppm at C-6. Chemical shifts and coupling constants for betaxanthins in the 1,7-diazaheptamethinium system are similar to betacyanins [5].

The O-glycosylation of the aglycone at 5 or 6 carbon is determined by difference of about 2 ppm downfield shift of the *ortho* proton and can be proven by NOE (Nuclear Overhauser Enhancement). The proton at C-11 was irradiated that enhanced the anomeric proton of the sugar, and 4-coumaroyl and feruloyl C-6′ esters of gomphrenin I were identified with the help of methyl proton in acyl moiety [33]; the same observation has been noted in lampranthin II (feruloylbetanin).

8.18 Chemical Assessment

Enzymatic or chemical methods have been normally used for structural elucidation of betalains in order to correlate the spectral data. The conformation of sugar linkage and the succeeding carbohydrate analyses could be suggested by the action with glucose oxidase or the hydrolysis of betalains with acid or enzyme (almond emulsion or β-glucosidase) might concede the sugar nature. To find the linkage, some more cumbrous chemical techniques are also applied [20]. By diazomethane methylation and following degradation with alkali giving 5-hydroxy-6-methoxyindole-2-carboxylic acid and 5-methoxy-6-hydroxyindole-2-carboxylic acid, position of the sugar is often scrutinized representing respective substitution at fifth and sixth carbon. For further characterization acyl group is released by alkaline hydrolysis in the case of acyl substituents. Position of acylation as well as the substitutions could be firmed chemically by permethylation, and succeeding acid hydrolysis give 3,4,6-tri-*O*-methyl-glucose and 2,3,4-tri-*O*-methyl-glucose in the case of C-20 or C-60 substitution, respectively. From treatment with b-glucuronidase and product analysis, the glucuronic acid configuration and existence could be implied.

In many circumstances, it is difficult to differentiate visually among anthocyanins. It is significant to note in plants of order Caryophyllales, betalains are typical plant pigments. For simple differentiation between betacyanins and anthocyanins, preliminary tests have been performed by means of the color displayed at different temperatures and pHs.

The most trustworthy way of differentiating among betacyanins and anthocyanins is electrophoresis by using acid buffer of 2–4 pH when the anthocyanins migrate as cations and betacyanins as anions. Acylated anthocyanins with dicarboxylic acid groups at pH 4 will also migrate as anions. Betaxanthins are likewise recognized by their anionic mobility, although yellow flavonoids, which are water-soluble, are immovable in these situations. When on a TLC plate coated with cellulose, an unidentified sample is smeared and developed in 1-butanol-acetic acid-water having 6:1:2 ratios for 2 h, betalain pigments will migrate somewhat slowly as related to anthocyanins. Inversion of these mobilities is done with a mobile phase having aqueous solvent having some acid [52].

Below is a table indicating the analysis parameters of betalain-producing species and families (Fig. 8.3).

Sources					
Family	Species	Pigments	Spectral characteristic MS peaks [M⁺]	Wavelength Λ_{max} (nm)	References
Amaranthaceae	*Beta vulgaris* L.	Betanidin(34)	389	541	[1]
		Betanin (35)	551	537	[2]
		2-Descarboxy-betanin (37)	507	532	[4]
		6′-o-Malonyl-2-descarboxy-betanin (38)	593	535	[4]
		Neobetanin[b] (51)	549	267,306,470	[11]
		Betalamic acid (1)	212	424	[12]
		Vulgaxanthin I (6)	340	470	[16]
		Vulgaxanthin II (7)	341	469	[17]
		ɣ-Aminobutyric acid-Bx (18)	297	459	[24]
		Serine-Bx (21)	299	468	[24]
		Valine-Bx (22)	311	470	[24]
		Phenylalanine-Bx (23)	359	472	[24]
		Isoleucine-Bx (24)	325	470	[24]
	Beta vulgaris var. lutea/*Beta vulgaris* L. SSP. Cicla [L.] Alef. Cv. Bright Light	Vulgaxanthin III (8)	326	470	[18]
		Vulgaxanthin IV (9)	325	470	[18]
	Beta vulgaris L. ssp. cicla [L.] Alef. cv. Bright Light	Alanine-Bx (25)	283	468	[26]
		Muscaaurin VII (26)	349	472	[26]
Mirabilis Jalapa	*Mirabilis jalapa* L./*Beta vulgaris* L.	Miraxanthin III (12)	331	473.5	[19]
		Miraxanthin V (14)	347	475.5	[21]
		Histamine-Bx (15)	305	468	[19]
	Mirabilis jalapa L./*Celosia argentea* var. cristata	3-Methoxy tyramine-Bx (13)	331	461	[20]

(continued)

(continued)

Sources			Spectral characteristic		
Family	Species	Pigments	MS peaks [M+]	Wavelength Λ_{max} (nm)	References
Portulaca grandiflora	*Portulaca grandiflora* (Hook)/*Opuntia ficus-indica* (L.) Mill	Portulacaxanthin I (3)	325,309	483	[14]
	Portulaca grandiflora/Beta vulgaris L.	Portulacaxanthin II (4)	375	468	[15]
		Portulacaxanthin III (5)	269	470	[15]
Opuntia	*Opuntia ficus-indica* L.	Indicaxanthin (2)	309	260,305,485	[13]
	Opuntia spp.	Phenethylamine-Bx (33)	315,270	475	[30]
	Opuntia sp.	Methionine-Bx (27)	343,299	477	[27]
Hylocereus	*Hylocereus polyrhizus*	Hylocerenin (46)	695	541	[9]
	Hylocereus ocamponis	4′-o-Malonyl-betanin (45)	619	538	[9]
		2′-o[5″-o-(E)-sinapoyl]-apiosyl-betanin (41)	889	330,550	[5]
		2′-o-Apiosyl-betanin (39)	683	539	[5]
Schlumbergera	Schlumbergera buckleyi	2′-o-Apiosyl-phyllocactin (43)	683	538	[8]
		2′-o-[5″-o-(E)-feruloyl]-apiosyl-phyllocactin (44)	945	328,549	[8]
Gomphrena globosa	*Gomphrena globosa*	Arginine-Bx (29)	368	469	[29]
		Lysine-Bx (30)	340,296	458	[29]
Carpobrotus	*Carpobrotus acinaciformis* L.	2-Descarboxy-betanidin (36)	345	533	[3]
Phytolacca	*Phytolacca Americana* L.	2'-O[5"-O-(E)-feruloyl]-apiosyl-betanin (40)	859	331,548	[6]
Phyllocactus	*Phyllocactus hybridus*	Phyllocactin (42)	637	539	[7]
Lampranthus sp.	*Lampranthus* sp./*Lampranthus sociorum* (L.Bol) N.E.Br. (MES.c.l.Bol)	Lampranthin II (48)	727	288,322,538	[10]

Sources					
Family	Species	Pigments	Spectral characteristic MS peaks [M^+]	Wavelength Λ_{max} (nm)	References
Bougainvillea	*Bougainvillea* sp.	Putrescine-Bx (32)	282	461	[28]
Celosia argentea	*Celosia argentea* var. cristata	Tryptophan-Bx (20)	398	218,264,471	[8]
Amaranthus tricolor	*Amaranthus tricolor* L.	Methylated arginine-Bx (19)	383	478	[25]
Rivina humilis	*Rivina humilis* L.	Humilixanthin (17)	325	258,463,4883	[23]
Glottiphyllum longum	*Glottiphyllum longum*	Dopaxanthin (16)	391	472	[22]

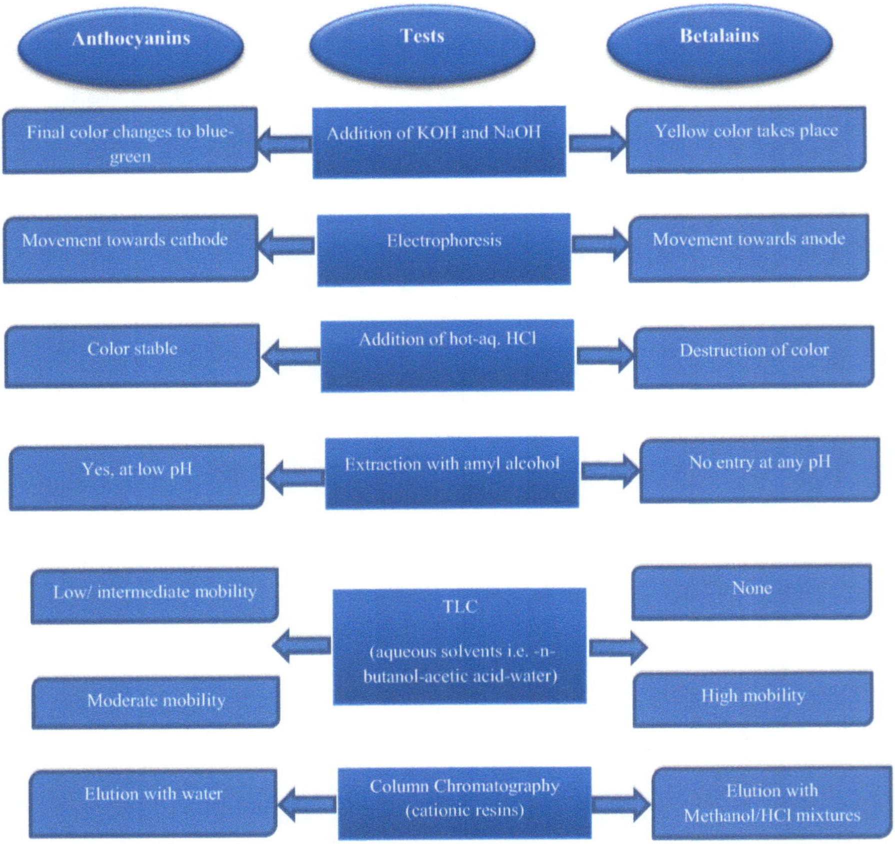

Fig. 8.3 Chemical assessments of anthocyanins and betalains

Many new identified betalains, as well as complete structures, have been determined by HPLC. However, structures of numerous reported betalains have yet to be fully revealed due to unavailability of commercial betalain standards. Isolation of standards from plant tissue is costly and time-consuming, and resulting standards differ in relative humidity, crystal water, and salts leading to underestimation or overestimation of molecules.

There exist ^{13}C data for only few molecules (e.g., neobetain).The use of more detailed spectroscopic techniques like NMR and MS for the definite documentation of betalains is needed. The new analytical approaches for betalain analyses like capillary electrophoresis and supercritical fluid chromatography are uncertain till yet and need to be explored in the future.

References

1. Wyler, H., Mabry, T. J., & Dreiding, A. S. (1963). U ber die konstitution des randenfarbstoffes betanin: zur struktur des betanidins. *Helvetica Chimica Acta, 46*, 1745–1748.
2. Piattelli, M., Minale, L., & Prota, G. (1964). Isolation, structure and absolute configuration of indicaxanthin. *Tetrahedron, 20*, 2325–2329.
3. Heuer, S., Richter, S., Metzger, J. W., Wray, V., Nimtz, M., & Strack, D. (1994). Betacyanins from bracts of Bougainvillea glabra. *Phytochemistry, 37*, 761–767.
4. Steglich, W., & Strack, D. (1990). Betalains. In A. Brossi (Ed.), *The alkaloids* (pp. 1–62). London: Academic.
5. Strack, D., Steglich, W., & Wray, V. (1993). *Betalains, in methods in plant biochemistry.* (Vol. 8, pp. 421–450). Orlando, FL: Conn, Ed. Academic Press.
6. Jackman, R. L., & Smith, J. L. (1996). Anthocyanins and betalains. In G. A. F. Hendry & J. D. Houghton (Eds.), *Natural FoodColorants* (pp. 244–310). New York: Chapman & Hall.
7. Delgado-Vargas, F., et al. (2000). Natural pigments: Carotenoids, anthocyanins, and betalains—Characteristics, biosynthesis, processing, and stability. *Critical Reviews in Food Science and Nutrition, 40*, 173–289.
8. Strack, D., Vogt, T., & Schliemann, W. (2003). Recent advances in betalain research. *Phytochemistry, 62*, 247–269.
9. Alard, D., Wray, V., Grotjahn, L., Reznik, H., & Strack, D. (1985). Neobetanin: Isolation and identification from Beta vulgaris. *Phytochemistry, 24*, 2383–2385.
10. Riaz, M. N. (1999). Soybeans as functional foods. *Cereal Foods World, 44*(2), 88–92.
11. Wohlpart, A., & Black, S. M. (1973). Accumulation of betanin in disks of Beta vulgaris leaves. *Phytochemistry, 12*(6), 1325–1329.
12. Sheu, M. J., & Wiley, R. C. (1983). Preconcentration of apple juice by reverse osmosis. *Journal of Food Science, 48*(2), 422–429.
13. Khan, M. I., & Giridhar, P. (2014). Enhanced chemical stability, chromatic properties and regeneration of betalains in *Rivina humilis* L. berry juice. *LWT-Food Science and Technology, 58*, 649–657.
14. Escribano, J., Cabanes, J., & Garcia, C. F. (1997). Characterisation of latent polyphenol oxidase in table beet: Effect of sodium dodecyl sulphate. *Journal of the Science of Food and Agriculture, 73*, 34.
15. Gandia, H. F., García, C. F., & Escribano, J. (2004). Purification and characterization of a latent polyphenol oxidase from beet root (*Beta vulgaris* L.). *Journal of Agricultural and Food Chemistry, 52*, 609.
16. Gandia, H. F., Escribano, J., & García, C. F. (2005). Characterization of the activity of tyrosinase on betaxanthins derived from (*R*)-amino acids. *Journal of Agricultural and Food Chemistry, 53*, 9207.
17. Stintzing, F. C., & Carle, R. (2004). Functional properties of anthocyanins and betalains in plants, food, and in human nutrition. *Trends in Food Science and Technology, 15*, 19.
18. Stintzing, F. C., Schieber, A., & Carle, R. (2000). Rote Bete als farbendes Lebensmittel —eine Bestandsaufnahme. *Obst Gem. Kartoffelver. Fruit Veg. Potato Process, 85*, 196.
19. Kugler, F., Stintzing, F. C., & Carle, R. (2004). Identification of betalains from petioles of differently colored Swiss chard (*Beta vulgaris* L. ssp. cicla [L.]Alef.cv. Bright Lights) by high-performance liquid chromatography–electrospray ionization mass spectrometry. *Journal of Agricultural and Food Chemistry, 52*, 2975.
20. Kobayashi, N., et al. (2001). Formation and occurrence of dopamine-derived betacyanins. *Phytochemistry, 56*, 429.
21. Schliemann, W., Cai, Y., Degenkolb, T., Schmidt, J., & Corke, H. (2001). Betalains of Celosia argentea. *Phytochemistry, 58*(1), 159–165.
22. Schliemann, W., Kobayashi, N., & Strack, D. (1999). The decisive step in betaxanthin biosynthesis is a spontaneous reaction. *Plant Physiology, 119*, 1217.

23. Steiner, U., Schliemann, W., Boehm, H., & Strack, D. (1999). Tyrosinase involved in betalain biosynthesis of higher plants. *Planta, 208*, 114–124.
24. Fossen, T., Slimestad, R., & Andersen, O. M. (2001). Anthocyanins from maize (*Zea mays*) and reed canarygrass (*Phalaris arundinaceae*). *Journal of Agricultural and Food Chemistry, 49*, 2318.
25. Stintzing, F. C., et al. (2002). A novel zwitterionic anthocyanin from evergreen blackberry (*Rubus laciniatus* Willd.). *Journal of Agricultural and Food Chemistry, 50*, 396.
26. Castellar, M. R., et al. (2006). The isolation and properties of a concentrated red-purple betacyanin food colourant from *Opuntia stricta* fruits. *Journal of the Science of Food and Agriculture, 86*, 122–128.
27. Barrera, F. A. G., et al. (1998). Stability of betalain extracted from Garambullo (*Myrtillocactus geometrizans*). *Food Science and Technology International, 4*, 115–120.
28. Azeredo, H. M. C. (2009). Betalains: Properties, sources, applications, and stability—A review. *International Journal of Food Science and Technology, 44*, 2365–2376.
29. Wyler, H. (1969). Die betalaine. *Chemie in unserer Zeit, 3*(5), 146–151.
30. Piattelli, M., & Minale, L. (1964). Pigments of centrospermae—II.: Distribution of betacyanins. *Phytochemistry, 3*(5), 547–557.
31. Scotter, M. J. (2011). Methods for the determination of European Union-permitted added natural colours in foods: A review. *Food Additives and Contaminants, Part A, 5*, 1–70.
32. Strack, D., Bokern, M., Marxen, N., & Wray, V. (1988). Feruloylbetanin from petals of *Lampranthus* and feruloylamaranthin from cell suspension cultures of *Chenopodium rubrum*. *Phytochemistry, 27*, 3529–3353.
33. Heuer, S., Wray, V., Metzger, J. W., & Strack, D. (1992). Betacyanins from flowers of *Gomphrena globosa*. *Phytochemistry, 31*, 1801–1807.
34. Huang, A. S., & Elbe, J. H. (1986). Stability comparison of two betacyanine pigments—Amaranthine and betanine. *Journal of Food Science, 51*(3), 670–674.
35. Adams, J. P., & Von Elbe, J. H. (1977). Betanine separation and quantitation by chromatography on gels. *Journal of Food Science, 42*, 410–414.
36. Steiner, U., Schliemann, W., & Strack, D. (1996). Assay for tyrosine hydroxylation activity of tyrosinase from betalain-forming plants and cell cultures. *Analytical Biochemistry, 238*, 72–75.
37. Wyler, H., et al. (1984). Cyclodopa glucoside ((2S)-5-(P- ~ –glucopyranosyloxy)-6-hydroxyindoline-2-carboxylic acid) and its occurrence in red beet (*Beta vulgaris* var. *rubva* L.). *Helvetica Chimica Acta, 67*, 1348–1355.
38. Wybraniec, S., Platznerb, I., Geresh, S., et al. (2001). Betacyanins from vine cactus *Hylocereus polyrhizus*. *Phytochemistry, 58*, 1209–1212.
39. Degenhardt , A., Engelhardt, U. H., Lakenbrink, C., & winterhalter, P. (2000). Preparative separation of polyphenols from tea by high-speed countercurrent chromatography. *Journal of Agricultural and Food Chemistry, 48*, 3425–3430.
40. Pourrat, A., et al. (1988). Betalains assay of fermented red beet root extract by high performance liquid chromatography. *Journal of Food Science, 53*, 294–295.
41. Chethana, S., et al. (2007). Aqueous two phase extraction for purification and concentration of betalains. *Journal of Food Engineering, 81*, 679–687.
42. Goncalves, L. C. P., de-Souza-Trassi, M. A., Lopeset, N. B., et al. (2012). A comparative study of the purification of betanin. *Food Chemistry, 131*, 231–238.
43. Wilcox, M. E., Wyler, H., Mabry, T. J., & Dreiding, A. S. (1965). *Helvetica Chimica Acta, 48*, 252–258.
44. Piattelli, M., Minale, L., & Prota, G. (1965). Pigments of Centrospermae—III.: Betaxanthins from Beta vulgaris L. *Phytochemistry, 4*(1), 121–125.
45. Piattelli, M., De Nicola, M. G., & Castrogiovanni, V. (1969). Photocontrol of amaranthin synthesis in Amaranthus tricolor. *Phytochemistry, 8*(4), 731–736.
46. Mabry, T. J., & Dreiding, A. S. (1968). The betalains. In T. J. Mabry, R. E. Alston, & V. C. Runeckles (Eds.), *Recent advances in Phytochemistry* (Vol. 1, pp. 145–169). New York: Appleton-Century-Crofts.

47. Trezzini, G. F., & Zryd, J. P. (1990). *Portulaca grandiflora*: A model system for study of the biochemistry and genetics of betalain synthesis. *Acta Horticulturae, 280*, 581–585.
48. Gandia-Herrero, F., Jimenez-Atienzar, M., Cabanes, J., Garcia-Carmona, F., & Escribano, J. (2010). Stabilization of the bioactive pigment of opuntia fruits through maltodextrin encapsulation. *Journal of Agricultural and Food Chemistry, 58*(19), 10646–10652.
49. Strack, D., Engel, U., & Wray, V. (1987). Neobetanin: A new natural plant constituent. *Phytochemistry, 26*, 2399–2400.
50. Wang, H., Cao, G., & Prior, R. L. (1997). Oxygen radical absorbing capacity of anthocyanins. *Journal of Agriculture and Food Chemistry, 45*, 304–309.
51. Schwartz, S. J., & Von Elbe, J. H. (1980). Quantitative determination of individual betacyanin pigments by high-performance liquid chromatography. *Journal of Agricultural and Food Chemistry, 28*(3), 540–543.
52. Tyman, J. H. P. (1997). The chemistry of some natural colourants. *Studies in Natural Products Chemistry, 20*, 719–788.
53. Schmidt, O. T., Becher, P., & Hübner, M. (1960). Zur Kenntnis der Farbstoffe der Roten Rübe, III. *European Journal of Inorganic Chemistry, 93*(6), 1296–1304.
54. Wyler, H., & Dreiding, A. S. (1959). Darstellung und Abbauprodukte des Betanidins. 3.(vorläufige) Mitteilung. Über die Konstitution des Randenfarbstoffes Betanin. *Helvetica Chimica Acta, 42*(5), 1699–1702.
55. Terradas, F., & Wyler, H. (1991). 2,3- and 4,5-secodopa, the biosynthetic inter-mediates generated from L-dopa by an enzyme system extracted from the fly agaric, *Amanita muscaria* L., and their spontaneous conversion to muscaflavin and betalamic acid, respectively, and betalains. *Helvetica Chimica Acta, 74*, 124–140.
56. Piattelli, M., & Minale, L. (1966). Structure of Amarantinand1soamarantin. 11. *Annali di Chimica, 56*, 1060–1064.
57. Minale, L., Piattelli, M., De Stefano, S., & Nicolaus, R. A. (1966). Pigments of centrospermae—VI.: Acylated betacyanins. *Phytochemistry, 5*(6), 1037–1052.
58. Strack, D., & Reznik, H. (1979). High-performance liquid chromatographic analysis of betaxanthins in Centrospermae (Caryophyllales). *Zeitschrift für Pflanzenphysiologie, 94*(2), 163–167.
59. Strack, D., Engel, U., & Reznik, H. (1981). High performance liquid chromatography of betalains and its application to pigment analysis in Aizoaceae and Cactaceae. *Zeitschrift für Pflanzenphysiologie, 101*(3), 215–222.
60. Gasztonyi, M. N., Daood, H., Hajos, M. T., & Biacs, P. (2001). Comparison of red beet (Beta vulgaris var conditiva) varieties on the basis of their pigment components. *Journal of the Science of Food and Agriculture, 81*(9), 932–933.
61. Forni, E., Polesello, A., Montefiori, D., & Maestrelli, A. (1992). High-performance liquid chromatographic analysis of the pigments of blood-red prickly pear (Opuntia ficus-indica). *Journal of Chromatography, 593*, 177–183.
62. Heuer, S., & Strack, D. (1992). Synthesis of betanin from betanidin and UDP-glucose by a protein preparation from cell suspension cultures of *Dorotheanthus bellidiformis* (Burm. f.) N.E.Br. *Planta, 186*, 626–628.
63. Bokern, M., & Strack, D. (1988). Synthesis of hydroxycinnamic acid esters of betacyanins via 1O-acylglucosides of hydroxycinnamic acids by protein preparations from cell syspension cultures of *Chenopodium rubrum* and petals of *Lampranthus sociorum. Planta, 174*, 101–105.
64. Bokern, M., Heuer, S., & Strack, D. (1992). Hydroxycinnamic acid transferases in the biosynthesis of acylated betacyanins: Purification and characterization from cell cultures of *Chenopodium rubrum* and occurrence in some other members of the Caryophyllales. *Botanica Acta: Journal of the German Botanical Society, 105*, 146–151.
65. Attoe, E. L., & Von Elbe, J. H. (1982). Degradation kinetics of betanin in solutions as influenced by oxygen. *Journal of Agricultural and Food Chemistry, 30*(4), 708–712.
66. Jackman, R. L., Smith-In, J. L., Hendry, G. A., & Houghton, J. D. (1992). *Natural food colorants* (Vol. 183). Glasgow, UK: Blackie.

67. Kujala, T. S., Vienola, M. S., Klika, K. D., Loponen, J. M., & Pihlaja, K. (2002). *European Food Research and Technology, 214*, 505.
68. Herrmann, K. (1977). Ubersicht iiber nichtessentielle Inhaltsstoffe der Gemiisearten I. Gurken, Melonen, Kiirbisse, Gemiisepaprika, Auberginen, Erbsen, Bohnen und Puffbohnen. *Z. Lebensmitt.-Untersuch, 165*, 87–98.
69. Nilsson, T. (1970). Studies into the pigments in beetroot (Beta vulgaris L. ssp. vulgaris var. rubra L.). *Lantbrukshogskolans annaler, 36*, 179–219.
70. Powrie, W. D., & Fennema, O. (1963). Electrophoretic separation of beet pigments. *Journal of Food Science, 28*(2), 214–220.
71. BE, J., Sy, S. H., MAING, I. Y., & Gabelman, W. H. (1972). Quantitative analysis of betacyanins in red table beets (Beta vulgaris). *Journal of Food Science, 37*(6), 932–934.
72. Chang, S. T., Wu, J. H., Wang, S. Y., Kang, P. L., Yang, N. S., & Shyur, L. F. (2001). Antioxidant activity of extracts from Acacia confusa bark and heartwood. *Journal of Agricultural and Food Chemistry, 49*(7), 3420–3424.
73. Barritt, B. H., & Torre, L. C. (1973). Cellulose thin-layer chromatography separation of Rubus fruit anthocyanins. *Journal of Chromatography A, 75*(1), 151–155.
74. Saguy, I., Kopelman, I. J., & Mizrahi, S. (1978). Thermal kinetic degradation of betanin and betalamic acid. *Journal of Agricultural and Food Chemistry, 26*(2), 360–362.
75. Wyler, H., Meuer, U., Bauer, J., & Stravs-Mombelli, L. (1984). CycloDOPA glucoside and its occurrence in red beet (Beta vulgaris var. rubra L.). *Helvetica Chimica Acta, 67*, 1348–1355.
76. Cai, Y., Sun, M., Wu, H., Huang, R., & Corke, H. (1998). Characterization and quantification of betacyanin pigments from diverse Amaranthus species. *Journal of Agriculture and Food Chemistry, 46*, 2063–2070.
77. Trezzini, G. F., & Zryd, J. -P. (1991a). Characterization of some natural and semi-synthetic betaxanthins. *Phytochemistry, 30*, 1901–1903.
78. Trezzini, G. F., & Zryd, J. -P. (1991b). Two betalains from Portulaca grandiflora. *Phytochemistry, 30*, 1897–1899.
79. Schwartz S. J., & von Elbe J. H. (1983). Kinetics of chlorophyll degradation to pyropheophytin in vegetables. *Journal of Food Science, 48*, 1303–1306.

Chapter 9
Bioavailability of Betalains

9.1 Introduction

Primitively the term bioavailability was used in materia medica in order to describe the rate and extent of action of a drug. Many definitions of bioavailability were recommended, but the more accepted one is the fractional part of any ingested compound that enters the systemic circulation and the particular sites where this compound maintains its biological action. Simply, it is quantitative measure of ingested betalains which exerts its advantageous biological activities in specific tissues [1].

Bioavailability is an important factor to determine the impact of a molecule on human health, i.e., whether the molecule is beneficial for health or not [2]. A food ingredient is beneficial to health if it is bioavailable *in vivo*. When food component is ingested, its active ingredient passes through the gastrointestinal pathway where it is absorbed, and its sufficient amounts are made accessible in the circulation, where the cells use them. Ingredient of food must be able to uphold its molecular structure in different digestion stages (that implies characteristic metabolism to the molecule through which its absorption extent and its expected rate are affected) in order to reach the systemic circulation and exert any health-promoting functions. Therefore any so-called health advantage of a food source should be affirmed firstly by bioavailability experiments that qualify *in vivo* absorption extent of food component. In this regard, betalains, the main bioactive part of *B. vulgaris* and inorganic nitrate bioavailabilities, are discussed in literature. Inorganic dietetically nitrate shows high bioavailability; it shows 100% absorption extent when it is digested. Betalains absorption extent is not sufficiently clear [3]. Different absorption extents are observed due to different parameters like source of food, molecular instability in the digestive tract, degradation due to bacteria in the intestine, and absorption mechanism [4]. Also, betalain absorption, metabolism, and removal are not fully described yet [5].

There are a few reports on metabolism and bioavailability of betalains [3]. Betalains oral bioaccessibility is considered low as compared to anthocyanins [5].

E. Akbar Hussain et al., *Betalains: Biomolecular Aspects*,
https://doi.org/10.1007/978-3-319-95624-4_9

Studies of bioavailability of betalains prove that betalains, betanin, and indicaxanthin are capable of residing inside the body, thus playing a role in health improvement by improving the redox state of the human body [6].

Bioavailability is influenced by various endo- as well as exogenous elements and variables, for example, configuration of sustenance grid, nourishment preparation, gulped mixes' science and stability, nutritional extent, collaborations with supplementary sustenance segments, bacteriological digestion system, and discrete contrasts. These could influence intestinal assimilation or dissemination in plasma and eventually the genuine phytochemicals grouping accessible to objective cells. Betacyanins are ingested and are distinguished in the pee of persons who consumed *B. vulgaris*, red beet juice [26, 30] or cactus pear fruit.

9.2 Bioavailability of Betalains from Red Beet

B. vulgaris is among a small number of vegetables that contain a group of highly bioactive pigments, betalains [3] (Fig. 9.1).

A bolus of *B. vulgaris* juice was ingested, and its presence in human pee was measured. After consumption of 300 mL *B. vulgaris* juice, Kanner et al. found 0.5–0.9% of the ingested betacyanins, i.e., betanin and isobetanin, in pee of volunteers after 12 h. This implies that betacyanins can easily be absorbed in humans but in small amount. They observed that the absorption significant peak of betacyanins' urinary excretion rate appeared after 2–4 h of ingestion; however, in this duration, there exists a high level of inter-individual unevenness [3]. After 3 h of ingestion, the observed concentrations were 0.2 mM betanin and 6.9 mM indicaxanthin in plasma. After ingestion of red beet juice, the range of bioavailability for betanin and isobetanin was observed to be 0.28–0.9% [7].

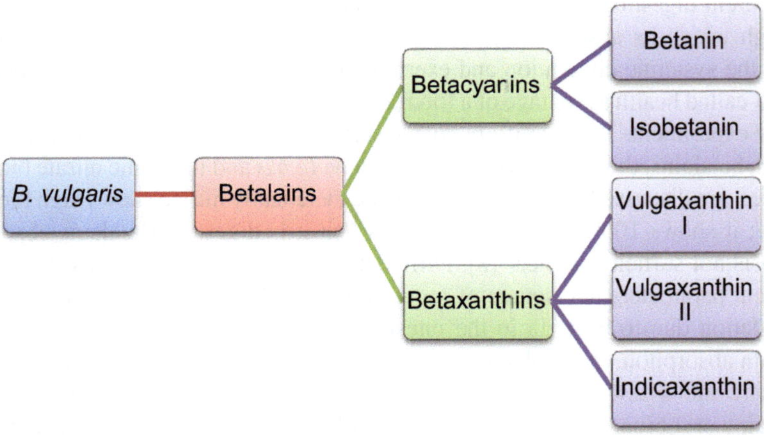

Fig. 9.1 Potentially bioactive betalains in B. vulgaris (beetroot)

Frank et al. also found the same results of bioavailability of betacyanins. He ingested 500 mL of *B.vulgaris* juice to six healthy volunteers and observed betacyanins in their pee after 24 h at concentrations equivalent to ~0.3% of the dosage ingested first. Small bioavailability level is revealed by these studies. It should be noted that through nephritic tract betacyanins are unlikely to be excreted solely [3, 9], and for absorption in the gut, hydrolysis is not required as indicated by the fact that these are absorbed in the systemic circulation in inviolate form from the gut and then these are released outside the body through pee [7]. In fact, urinary elimination cannot be used as an exclusive bioavailability indicant because in bilious and circulatory pathways, clearance of ingested ingredient is not apparent through it; therefore bioavailability is not thoroughly studied. While studying bioavailability, betalains metabolic extent and its structural transmutation to secondary metabolites should be taken into account which needs to be explored completely [3].

Regardless of numerous researches on betalains' natural action, the course of betalain assimilation in the body is still not clear. Investigations of betanin bioavailability are summarized in Tables 9.1 and 9.2. *In vitro* studies suggested that the assimilation of betanin happens in the digestive system. When fed to rats, around 2.7% unmetabolized pigment excretion happened in defecation and pee independently. It was expected that the greater part of ingested betanin was absorbed in the gastrointestinal tract. Saline suspensions of stomach divider metabolize about 74% betanin, whereas the small digestive tract and colon metabolize 35% and 60% betanin, respectively. On the other hand, the liver metabolizes only a little amount of betanin as demonstrated by its perfusion tests. A later report confirmed the presence in pee; however, the degradation in the colon, small intestine, and stomach was similarly less, most likely because of the way the example is utilized (extract that is enriched in betalain). When betalain extracts of hepatocytes were incubated for 7 h, no metabolic impacts were identified, connoting that the liver is not a positive place for metabolism of betalain. The analytical process might neglect to recognize its metabolic items. Case in point, amid spectrophotometric investigation of the recouped pee 24 h in the wake of sustaining betalain extract, aside from in place betanin, no applicable metabolic item was identified. Likewise, spectrophotometric examination of pee of humans' candy stripers, 0.5–0.9% bioavailability of betanin was revealed. Two to four hours after ingestion, the pee was colored that was preceded up to 12 h, where no betalain metabolism was distinguished. By comparison of quercetin glucoside digestion procedure in the gut, it was guessed that the ingestion site is the digestive system. In 2004, Tesoriere et al. introduced HPLC broke down information on plasma and urinary rate of excretion of human volunteers who were nourished 500 g of cactus pear natural product pulp containing betanin in 16 mg amount over 12 h' time. In plasma, betanin showed up after 1 h, and C_{max}(extremum plasma concentration) of 0.20 nmol/mL was observed after 3.1 h (T_{max}). For betanin, the fatal disposal half-life ($T_{1/2}$) was 0.94 h. Lower half-life time ($T_{1/2}$) for betanin, contrasted with indicaxanthin, recommended the likelihood of lower dispersion volume, higher clearance rate, or both of these. In this study, betanidin was not observed that suggested that retention of betanin is independent of deglucosylation. Variety in bioavailability of betanin in pee in light of nutritional

Table 9.1 RDA values for betalain

Principle	Energy	Carbo hydrates	Protein	Total fat	Cholesterol	Dietary fiber	*Phytonutrients*	Carotene-ß	Betaine	Lutein-zeaxanthin
Nutrient value	43 cal	9.56 g	1.61 g	0.17 g	0 mg	2.80 g		20 µg	128.7 mg	0 µg
RDA percentage	2%	7%	1%	0.5%	0%	7%		—	—	—
Vitamins										
Principle	Folates	Thiamin	Pantothenic acid	Pyridoxine	Riboflavin	Niacin	Vitamin A	Vitamin C	Vitamin E	Vitamin K
Nutrient value	109 µg	0.031 mg	0.155 mg	0.067 mg	0.057 mg	0.334 mg	33 IU	4.9 mg	0.04 mg	0.2 µg
RDA percentage	2%	2.5%	3%	5%	4%	27%	1%	8%	0.5%	0%
Minerals							*Electrolytes*			
Principle	Ca	Cu	Fe	Mg	Mn	Zn			Na	K
Nutrient value	16 mg	0.075 mg	0.80 mg	23 mg	0.329 mg	0.35 mg			78 mg	325 mg
RDA percentage	1.5%	8%	10%	6%	14%	3%			5%	7%

Source: USDA National Nutrient data base

Table 9.2 Betalains' estimated yearly fabrication potential extracted from eatable origins

Sources (plant part)	Yield (t/ha)	Betalain content (mg/g fw)	Global production of plant (kt)	Total betalain yield[a] (t)	References
Red beet (*B. vulgaris*) Swiss chard[b] (petiole)	50–70 35–40	0.4–20 0.04–0.08	241,985,317,000 –	96,794,126,800 –	Pavokovic et al. [30] and DAFF (n.d.) Stintzing et al. [31]
Amaranth[c] (seed)	0.001	0.5–1.99	0.05	0.02	Cai et al. [22]
Cactus pear (fruit)	45	0.8	23.4	18.7	Guevara et al. [32]
Pitaya (fruit)	7–19[d]	0.32–0.41	194.7[e]	60.3	Bellec et al. [33], and Vaillant et al. [34]

[a]Whole content of betalain was designed taking into account the inferior values of its contents in respective origins
[b]Almost 50% postharvest loss of Swiss chard petioles renders its noneconomical source.
[c]Amaranth's global production numeral is grounded on yearly production in America, Africa, and other countries in 2005. A 100,000 acre in China is producing it. Pigment can be extracted from almost all parts of the plant involving stem, root, leaves, seed, and inflorescence
[d]Average yield in the USA and Malaysia
[e]Its worldwide production approximation is grounded on the import of Europe, Vietnam, and Malaysian fruit production in 2005. It also comprises the America's import of 1200 t pitaya fruit from Nicaragua.

cause was evident as betanin obtained from fruit of cactus pear was principally more bioavailable (3.7%) than that of red beet. When pee of human candy stripers nourished with red beet juice was examined via LC-MS/MS and spectrophotometric analysis, just 0.28% betanin bioavailability was uncovered, which was practically identical to that of anthocyanins. These qualities, particularly the bioavailability and $T_{1/2}$ (7.43 h), were conflicting with information obtained from pharmacokinetics of plasma because Frank et al. [35] did not explain this. The variety in information could not be accurately represented. Furthermore, betalains' metabolic change results were neglected, even if some anonymous crests were seen in the HPLC analysis results of pee tests. In this way, metabolic change results of betalains were evading identification by specialists, and, consequently, bioavailability concentrates so far are not comprehensive. Additionally, bioavailability referred to in this segment seeming to have disparities in one or more determinants of the investigations that ought to be dealt with, for example, testing span, handling of test, analytical technique, and resulting betalains' measurement. Taking into account the poor stability of betalains, different examining span and test preparations might influence the outcomes. Conflicting results could likewise be somewhat ascribed to various scientific techniques that show absorbance at 535–538 nm and consequent measurement utilizing related termination coefficients, 60,000 and 65,000 L/mol.cm. The combined impact of the varieties in the investigations might be in charge of the irregularity in betanin bioavailability facts, uncommonly from juice of red beet. By and by, it is irrefutable that betanin exhibits low bioavailability. While trying to comprehend the purpose behind constrained betanin bioavailability, betanin transferal that is transepithelial was examined in cells of Caco-2. The outcomes gave additional confirmation of assimilation of betanin through epithelial cell lining of intestines with no metabolic changes. Betanin from red beet was not as much retained in contrast to cactus pear, most likely because of grid impact. Betanin displayed a permeability coefficient which demonstrated critical bidirectional qualities and nonlinear outflow energy. Several other confirmations also recommended that retention of betanin was specifically blocked by a protein 2 that acts as multidrug resistance (MRP2)-interceded effluence and also the foodstuff medium. Then again, betanin's poor bioavailability could likewise be ascribed to debasement of the shade in the gastrointestinal tract and post absorptive circulation in whole body of organism. As to betanin content in the gastrointestinal tract, absorbability including dismembered bits of the small digestive system demonstrated that 26–60% of betalains, particularly betanin or betacyanins, were metabolized essentially showing a conceivable purpose behind pitiable bioavailability. So also, in vitro recreation investigations of betanin degradation at little intestinal, gastric, and oral stages uncovered around 50% loss of pigment. Additionally, bioaccessibility of betanin was constrained by sustenance network, which may be a purpose behind low, and differential, bioavailability relying upon dietary source. Moreover, the little bioavailability could likewise be credited to postabsorptive dissemination of betanin in various compartments of the body, for example, erythrocytes [2].

Bioaccessibility of indicaxanthin is more significantly accounted for as compared to betanin in human candy stripers (Table 9.1). Seventy-six percent of indicaxanthin was ingested by human volunteers who were nourished with 500 g fruit pulp of cactus pear consisting of 28 mg indicaxanthin, as observed by HPLC analysis. Likewise, the discharged betalains' molar proportion was tantamount to the proportion of the AUC0–12 of betanin and indicaxanthin. This could be a direct result of comparative metabolic routes of betanin and indicaxanthin. In plasma, indicaxanthin was distinguished after 1 h, revealing 6.9 nmol/mLC_{max}. The T_{max} was noted as 3 h, while $T_{1/2}$ was noted as 2.36 h. High bioaccessibility of indicaxanthin was bolstered by pharmacokinetic approach reportage with a solitary oral organization of 2 μmol/L indicaxanthin per kg body mass to rats. It was also revealed that plasma grouping of indicaxanthin ($n = 15$) at 0.22 μmol/L, exhibiting 1.15 h half-life time, which was partial of the worth described beforehand in human candy stripers. Besides, in pee of rat, indicaxanthins' bioavailability was just 21% against 76% in human volunteers. The distinction in C_{max} of plasma and renal discharge in rat and human studies might be because of clear contrasts in the rat and human frameworks as watched for phenolic mixes. Not just this, the intellect could likewise be the wellspring of indicaxanthin: purified indicaxanthin and fresh pulp of fruit, which might be extra, inclined to oxidation in the gastrointestinal tract. All the studies were in complete agreement with the previous ones indicating the small metabolism of betalains in the liver. Despite the fact that indicaxanthin assimilation by means of epithelial cells of intestines was proposed before, it was exhibited in cells of Caco-2. Further, trying to quantify the components deciding the diffusion of indicaxanthin, carriers that depend upon proton, transporters of efflux, concentration of pigment, and coefficient of permeability were examined. The study indicated contribution of bearers and transporters in indicaxanthin convergence or efflux, dispersion coefficient of permeability was made non-polarized, that depends upon concentration and time, and is not affected by membrane transport inhibitors. Indicaxanthin ingestion in erythrocytes is accounted for, maybe, through paracellular dispersion, independent of carriers, as on account of phenolic acids. In the gastrointestinal tract, it creates the impression that indicaxanthin is verging on getting away from the debasement process. In vitro reenactment involving intestinal, gastric, and oral phases revealed the loss of only a minute quantity of indicaxanthin in the whole process of digestion. Furthermore, indicaxanthin was not affected by nourishment lattice and remained completely bioaccessible. In any case, privilege of indicaxanthin's finished avoidance of phases of digestion needs advance confirmation as prior absorbability tests including analyzed bits of stomach, small digestive tract, and internal organ recommended critical. A few individuals pass red pee, even in the wake of eating a humble measure of *B. vulgaris*. This happens on the grounds that the color betanin goes conventionally via their digestive framework and is innocuously discharged in stools and pee. This disorder is referred to as betacyaninuria or beeturia. Betanin are generally consumed by the gastrointestinal framework thus don't ordinarily shading the pee so drastically.

No drastic impacts are caused by beeturia but may come with unpretentious outcomes in the event that sometimes it may be misdiagnosed by hematuria (blood in the pee). In 1836, Dr. Rees presented an article in *The Lancet* red pigmentation was observed to produce from vegetable matter, as a patient also told that had taken *B. vulgaris* in salad. A false judgment of hematuria can start pointless and dangerous medicines. Nowadays, transfer of doubted hematuria can happen that ends up being beeturia.

It has for quite some time been realize that a few people produce pink or red pee in the wake of *B. vulgaris* intake, whereas others seem insusceptible to the disorder. This perception drove researchers in the 1950s to infer that it is a hereditary quality, that is, a picked up disorder that has gone down over and done with the eras. Generally, it is said that beeturia happens to a single individual in every eight. In 1950, Zindler and his colleague described a figure in a specific order after analyzing 78 different patients; they found 12.6% cases affected with beeturia. Allison and his colleague in 1956 described that out of their 104 subjects, 9.6% (10 subjects) suffered from beeturia. They declared that the characteristic is controlled by a particular recessive gene. Notwithstanding, different researchers, incorporating Penrose in a memo to the *British Medical Journal*, were incredulous of these researches, which achieved deductions in view of lacking information. A generally described numeral for beeturia is that it happens in 14% of the population. This was based on the research conducted by Watson et al. at the Glasgow Royal Infirmary, distributed in 1963. A premature hypothesis of beeturia states that it is connected with nutrients distaste. The group at Glasgow recorded an observation that nutrients distaste was not a variable in beeturia creation. The gathering, be that as it may, provided a profitable understanding into the disorder by taking a gander at beeturia in groups of individuals possessing various medical circumstances. 13.8% beeturia rate was observed in control group, but the other medical group came forth with high rates of beeturia due to iron insufficiency. That iron-inadequate subject of medical observation had red-shaded pee in about 80% range. Inadequacy of iron in subject group was brought about by a few components, such as hemorrhoids, hernias, draining peptic ulcers, as well as poor diet. Supplementary testing was done after iron supply to subject that had deficiency of iron, alongside a standard extent of *B. vulgaris*; 49% rate of beeturia was reduced.

This research at Glasgow began a change from considering beeturia as a principal hereditary disorder, in the direction of one that is resoluted ecologically. In 2001, Steve Mitchell who was from London, Imperial College, assessed all the accessible information on beeturia. He presumed that mostly it is a role of physical composition of a person which ruled out the earliest assumption that it is caused due to direct polymorphic hereditary switch. A scope of confirmation has toppled the basic hereditary model for this disorder. At first, individuals having this disorder did not have the reliable condition.

The red pee excretion goes back and forth after some period, proposing extra natural components. A 1990s research exhibited that people who have the same intake the same measure of *B. vulgaris* at various distinctive events demonstrated an extensive variety of difference in the measure of betanin discharged in their pee. Besides, as investigations of hale and hearty people were collected, there existed an undeniably extensive variability of variety in extent of individuals answered to suffer from beeturia. Testified all the subjects discharged red pee in the wake of devouring *B. vulgaris*. Apparently, some undisclosed natural variable in this populace was bringing on the abnormal amounts of this disorder. In a third place, spectrophotometric utilization in current beeturia contemplates demonstrated that there exists a constant range in the redness level of pee inside of one populace. On the off chance, the complaint was transferred hereditarily to distinctive classes like excretors and non-excretors. In these researches, the non-excretors might basically be delivering a very light shade that can be seen by naked eye. Just like 1960s, the expanding affectability of the systematic apparatus utilized might likewise clarify the advanced note down frequency of beeturia in these researches. In the fourth place, contemplates with identical twin have neglected to demonstrate any conspicuous genetic example for this disorder. In final, the utilized *B. vulgaris* type influenced the result as a part of one beeturia study, with people giving seriously hued pee in the wake of expending one cultivar, however a typical pee shading in the wake of eating an alternate one. Betalain color substance is known not among *B. vulgaris* cultivations. Steve Mitchell took note of that season of cultivation and collected *B. vulgaris*, while some pre-bundled varieties improved *B. vulgaris* accumulation so as to shade concentrated *B. vulgaris* extricate. Besides, chemicals present in *B. vulgaris*, for example, ascorbic and oxalic acids, can restrain the gastrointestinal framework's capacity for betalain synthesis. All things considered, these variables constitute the extensive variety in the described beeturia rate. In the stomach, due to the bacterial activity and acids, degradation of betanin occurs to a specific extent; however, the color is for the most part ingested in the large intestine or colon. Ingested betanin is transported through the intestinal divider via an acceptor substance; along these lines little is recouped from the pee. Conversely, if infusion of betanin takes place into the circulation system, avoiding the framework of the digestive system, the vast majority of it is recuperated in the pee. In beeturia's instances, the process of ingestion in the digestive system gets disturbed. Insufficiency of iron has all the earmarks of being the best interrupter of assimilation of betanin. The oral organization of iron can in part redress the condition. There exists no straightforward relationship between beeturia and level of iron in the blood. In the event that huge amounts of *B. vulgaris* are ingested, it ought to be noticed; the shade will overburden the digestive framework's ability to ingest it, and red pee will more likely than not be discharged [3].

9.3 Bioavailability of Betalains from Cactus Pear

Bioavailability of betalains is studied by its plasma kinetics and excretion through pee of healthy volunteers by ingestion of single bolus; 500 g of *Opuntia ficus-indica* (cactus pear) fruit pulp comprise of 28 mg of indicaxanthin and 16 mg of betanin. After 3 h of ingestion, concentrations of phytochemicals betanin and indicaxanthin were observed through plasma peak. Within 12 h, both phytochemicals vanish from plasma [4]. The bioavailability of indicaxanthin (7 mmole/L) was 20 times high as compared to betanin (0.2 mmole/L) [8].

For both compounds, non-exponential plasma downslope was observed which reveals that tissue storage was nonselective. After 12 h, pee recovery of indicaxanthin was 75%, and that of betanin was 3.5% of the ingested material. Cactus pear is a meliorate source of bioavailable betanin as compared to red beet through which bioavailability is no more than 1%. In human beings, after intake of 500 g fruit pulp, red blood cells and LDL (low-density lipoproteins) were analyzed for betalain post-absorption distribution. In RBCs and LDL, recovery of both phytochemicals was done to an extent that seemed to be a reflection of their plasma. Betalains are amphiphilic due to which these can bind easily to LDL particles as well as it is responsible for convenient division between plasma and RBCs [4].

In order to explore bioavailability of betalains, a study advanced than the one described above was done by using both cactus pear and red beet juice. Tesoriere et al. in 2008 conducted a study feigning oral, small intestinal, and stomach digestion; evaluation of digestive stability of purified pigments was done. Indicaxanthin loss was 24B (minor), and there was a decrease of vulgaxanthin I in all steps of digestion. Matrix of food showed no effect. In contrast, it prevented betanin and isobetanin decay in an imitated gastric surroundings. From food samples and purified samples, betacyanin obtained was able to trace in the small intestinal phase of digestion imitation. In purified pigments case, betalamic acid was assembled after digestion, while in food containing betalains case, no assemblage was observed. In the post-intestinal digestive part after ultracentrifugation, betaxanthins were dissolved completely in the bioaccessible aqueous fraction. However, from the matrix of food, an incomplete ejection of betacyanins takes place. In post-intestinal digest case, an oxidative suppression of methyl linoleate in methanol was detected which seemed to depend upon dose, although such impressions could not be mutually related with the betalain's level. Therefore in the digestive tract, dietary betalain stability is connected directly with its bioavailability, although the effects of some other factors such as the food matrix and processing style may also affect betalain bioavailability [7].

In spite of the fact that pigments of biological inception are extremely investigated for their arcade aptitudes and human health fitness, great trouble is encountered in their stability issues, from the ripeness of the origin to their handing out and food stuffs in which incorporation of pigments is done and preserving those products.

Usage of betalains like natural pigments is influenced by intrinsic as well as extrinsic elements. In spite of the fact that the development of the *B. vulgaris* is not a significant quality distinctive in sourcing, different preparation stages, including the conditions of extraction, extraction medium determination, item concentration, capacity, joining, and transportation into items, are immense undertakings. Stability of betalains is influenced by the extrinsic variables that tend to incorporate oxygen, pH, temperature, light and chemicals of handling, and nutritious matter. Contrastively, the intrinsic factors are basically enzymes: β-glucosidases, peroxidases (PODs), and polyphenol oxidases (PPOs) aside from the nature of the origin of red beet.

In addition, a number of other factors such as chemical changes occurring at degradation step and their supplementary fortune also influence the stability of betalains. Upgrades in storage conditions as well as extraction or handling strategies that confer stability to betalain pigments have been examined. Moreover, use of propelled procedures such as fermentation utilized for the pigments planning and preparations of food are also investigated by researchers.

The *M. geometrizans* (garambullo tree) which develops in the deserts of Mexico creates a purple natural product. The pigments were extricated, recognized, and assessed for their stability at distinctive temperatures and pH values and in the vicinity of iron and chromium. One percent citrus extract or ascorbic acid was included as a stabilizer, and a blend of both was likewise utilized. On the premise of their noticeable light spectrum and chromatographic profile, the pigments were recognized as betalains, which have more prominent stability than that of red beet colors and which are exceptionally steady at low temperatures.

Betalains biosynthesized by *B. vulgaris* (beetroot) bristly root culture were tested for its radical searching movement and stability at distinctive pH range, concentration of bile salts, and also the in vitro gastrointestinal states. It was noted that stability of betalains remains unaffected at 4% concentrations of bile salt and $pH < 3$. Betalains are moderately stable at the in vitro states of the gastrointestinal tract, as is obvious by their radical scavenging activity that diminishes from 75% restraint of 2,2-diphenyl-1-picrylhydrazyl (DPPH*) to around 38%.

Betalain pigment also known as purple pitaya from *H. polyrhizus* (red dragon fruit) was extricated, and its stability was assessed. Homogenization of natural products was done with ethanol to discrete pectic materials. UV/Vis spectrophotometric analysis demonstrated absorbance peaks at 230 and 537 nm of its fruit extract. Vicinity of betacyanin is demonstrated by absorbance peak at 537 nm. The progressions of betacyanin intensity because of light, pH, temperature, storage condition, and added substances were checked for 21 days at 537 nm with an UV/Vis spectrophotometer. Outcomes exposed light as the central point of degradation of betalain pigment. Storage at 4 °C in refrigerator without presentation of light figured out how to protect the shade of natural product juice up to 21 days (3 weeks).

9.4 Uses and Commercial Production of Betalains

Chlorophylls, carotenoids, anthocyanins, and betalains speak of the four most essential color classes conferring a wide range of color shades to foods grown from the ground. These colored pigments are considered to advance human health and well-being.

Later on, extra examinations will be required to comprehend parts of the physiology of betalains in the human body, conceivable chemo-preventive activities, and advantages of an eating routine rich in betalain pigments.

As of late, there has been expanding enthusiasm for betalains since a few studies have indicated their conceivable antioxidant effects. In the event that these examinations are affirmed, the additional commercial estimation of these shades will be expanded because of the developing utilization of antioxidants in the sustenance industry for their helpfulness in conservation as well as a result of their advantageous consequences for human health.

Subsequent researches in the assortments of Asian and European cactus pears have demonstrated outstanding antioxidant potential that fundamentally diminish patients' oxidative stress and might prove beneficial counteracting constant deviations from healthy conditions. Betalain pigments of these cactus pears indicated helpful impacts on the oxidation-reduction directed routes included in development of cell and inflammation. A few researches have recommended that polyphenolics, vitamin C, and flavonoid mixes (i.e., quercetin, isorhamnetin, and kaempferol) in cactus pear are responsible for its aggregate antioxidant activity agent movement. Betalain pigments are soluble in water. Two betalain subordinates such as betacyanin (red-purple) and betaxanthin (yellow-orange) are present in cactus pears. These colors exhibit imperative antioxidant without harmful impacts in people. Likewise, flavonoids of cactus pear are responsible for neuroprotectory movement against oxidative harm prompted in societies of cortical cells of rat. Additional advantageous impact of the fruit was seen in the counteractive action of ulcers of stomach through the incitement of generation of prostaglandin: cactus pear advanced mucous discharge of bicarbonate, included in the assurance of mucosa of gastric regions (F6). They can be utilized as a part of the treatment of inflammatory and cardiovascular illnesses, cancer, asthma, arthritis, oxidative stress, intestinal aggravation, diabetes, and different sicknesses connected with aging. The additional estimation of these shades is expanded inferable from their twofold capacity as colorant and as antioxidant.

Betalains have a few applications in sustenance, for example, sweets, dry mixtures, meat, and dairy items. The pure pigment concentration compulsory to get the desired shade is moderately insignificant, once in a while surpassing 50 mg, 1 kg figured as betanin. In 2004, Codex Alimentarius Commission caused restriction of betalain usage by Good Manufacturing Practice. Red beetroot food colorant, removed from *B. vulgaris*, is marketed in the USA and European Union as nourishment colorant. The leaves and stems of youthful plants are steamed quickly and eaten as a vegetable; more established leaves and stems are stir-fried. The typically dark red roots of garden beet are eaten either as a cooked vegetable or cold as a serving of salad in the wake of cooking and including oil and vinegar. An extensive extent of commercial production is prepared into boiled and sterilized beets or into pickles. In Eastern Europe beet soup, for example, cool soup, is a prominent dish. Small-scale development of yellow-hued garden beets is done for home utilization. *B. vulgaris* can be eaten warm with margarine as a delicacy after peeling and steaming or chilly as a fixing after cooking and salting or served as salad after peeling and destroying raw. Cured beets are a conventional sustenance of the American South. *B. vulgaris* color is utilized industrially as a nourishment color. It changes color when warmed; so it must be utilized as a part of desserts, ice-cream, and other sweets, yet it has no known allergic side effects, and it is economical. *B. vulgaris*, is a typical serving of salad ingredient when baked. *B. vulgaris* juice is exceptionally intense, and it's suggested that you drink the raw juice diluted no less than four times with other milder squeezes, for example, carrot and cucumber. There are nine different species in the *Beta* genus, and all additionally have the normal name beet, with *Beta vulgaris* being the most surely understood and commercially imperative making it known as the basic beet. Beets, with substantial leaves, are additionally developed as decorative plants. Biologically, they give sustenance to numerous creatures, including the larvae of various nuisance species. Beets are delightful for their shading and flavor and in addition for their beet nourishment. Their juice is magnificent blended with carrot juice and can likewise be utilized as a color. In a few nations, the beet juice, betanin, is handled commercially for shading in different items.

Commercial availability of beet pigments is possible as powder containing 0.3–1% of pigment or as juice concentrates created by beet juice vacuum concentration to 60–65% aggregate solids [1]. Cai and Corke analyzed betacyanins from *Amaranthus* and commercial pigments as far as their color qualities and stability at various temperatures in model sustenance frameworks. As compared to red radish anthocyanins, betacyanins displayed a brighter-red color. Both colors indicated comparable shading solidness at 14 and 25 °C; however, at 37 °C, the color of betacyanin was not as much of steady as red radish anthocyanin. An engineered colorant was additionally tried, which was steadier than betacyanins in utmost stockpiling circumstances [2].

Research demonstrates betalains are huge cancer prevention agents with mitigating properties that may shield human cells from oxidative stress. Betalains, phytonutrients found in beets, desert plants, Swiss chard, and blooming amaranth plants, may likewise address untimely aging and detoxify the body. These colors range from yellow to dim purple and are utilized in the food industry as a dye.

Two dozen types of betalains exist in the Caryophyllales plant family, with two primary structures contemplated for their medical advantages. Betacyanins speak to the color utilized as a part of nourishment colors, which show up as red, blue, and purple colors [10].

Betaxanthins represent to a yellow pigment likewise found in beets yet at lower levels than the red assortments. Cacti developed in a few ranges, for example, Sicily, contain adequate measures of these mixes. The Sonora cacti nopal prickly plants found in Mexico can persevere through amazing temperature varieties, which permits them to deliver every 1 of the 24 betalains. These cacti have been utilized as nourishment and drug by local individuals for a huge number of years.

Contemplates directed with the bright, pink product of nopal desert flora, likewise called thorny pear, show betalains may fight poisons in the human body that cause cell harm. Individuals are presented to poisons day by day through the air, water, and some sustenance. These free radicals have been connected to tumor and different maladies. Analysts trust supplements in the prickly plants may secure against tumor development, particularly in the mind.

Betalains may likewise lessen aggravation brought on by contamination. They may likewise bring down levels of hurtful cholesterol in the blood that prompts the development of plaque and blocked arteries [11].

Betalains may help lower levels of harmful cholesterol that may cause blocked arteries Some well-being nourishment stores offer a beverage produced using nopal cacti juice blended with fruit nectar to enhance flavor. The beverage is sold as a supplement to detoxify the body and repair cell harm brought on by free radicals and different poisons. Betalains may flush these poisons from the body, particularly the liver [12].

Betalains may be helpful in flushing toxins from the liver Nopalea, the trademark name for this well-being beverage, is likewise advanced as an antiaging product. Makers claim betalains in the juice repair harmed skin and may avoid untimely aging. Nopalea is likewise advertised to assemble a solid insusceptible framework since it may kill free radicals. The well-being beverage may likewise build vitality levels [13].

Commercial production By Vargas et al. [14], the commercial creation of common pigments has been lingering behind because of deficiency of critical amounts of fresh tissues of the plant that are very much pigmented, absence of straightforward and effective extraction strategies, and refinement of these items. Investigation is completed on only 30% of all pigmented plants; 0.5% of it has been comprehensively concentrated on. For instance, carotenoids creation by plants, green growth, and dinoflagellates has been evaluated at a hundred million ton every year [15]. In any case, toward the end of twentieth century, carotenoids established just a section of the normal color market. Assessed capability of yearly worldwide betalains creation positions at 96.8 Gt, of which *B. vulgaris* guarantees around 99.99%, while 79.1 t comes from different sources such as cactus pear, pitaya, and amaranth. On the off chance that 0.5% of the yearly generation of the plant parts comprising of betalain pigments is used for extraction of color, all generation limit of pigment would be 0.48 Tt every year [16]. Demonstrating much lower commercial generation of betalains in EU, crude or refined beet separate represented ca. 4000 t as it were. Various other betalain sources are not known. A little extraction level betalain at commercial scale is mainly due to the less stability of betalains bringing about critical misfortune amid handling, and just extraction of betacyanins is monetarily suitable as betaxanthins loss during extraction process is similarly greater as compared to betacyanins [17]. This plainly demonstrates the requirement for mechanical change in extraction and ensuing preparation to minimize color loss. As of late, endeavors on investigating new and monetarily practical sources of normal hues have brought about expansion of numerous new sources and mechanical change in extraction, handling, and application. Subsequently, worldwide commercial sector for nourishment hues developed by 2.1% amid 2009 to achieve US$ 1.45 billion. In 2009, the worldwide interest for sustenance pigment was 45,000–50,000 ton, and it might be expected that ca. 10% of the interest was satisfied by supply of betalains that is obtained from *B. vulgaris*. Anyhow, significantly, more mechanical developments are needed in the future to endeavor more natural origins for extraction of betalain [18]. Betalains' biotechnological production through in vitro frameworks could convey the commercial sector request attractively; however, mass proliferation of plants could be likewise truly considered in light of the fact that few plants are easy but difficult to keep up and develop on dry barren area. After the color has been extracted, some fruits such as cactus fruits could be utilized as wellspring of critical phytochemicals after color extraction, and the parts of plant could likewise be used viably as food of animals. Some recently discovered novel betalain sources such as *U. tuberosus*, *S. salsa*, *B.alba*, *C.formosanum*, *R. humilis*, and *T.triangulare* give off an impression of being promising for the most part because of the other gainful phytochemicals stated in them. Some plants such as pigeon berry, pokeweed, normal and feathered cockscomb (*Amaranthus* sp.), cactus pear (*O. ficus-indica*), and bougainvillea have the capability of supplanting red beet as betalains origin. All the while, a couple of different points of interest, for example, accessibility of wide range of color, better nature of colors without off-flavor, lessened nourishment versus colorant struggle, and probability of creating betalain-invigorated plans comprising some other biologically active and healthful segments, could be normal [19].

9.5 Production of Betalain as Colorants

Before biotechnological generation of betalains, production of colors was economically unfeasible. Economic contemplations incorporate high introductory yield of color, enhancement of the procedure of extraction, and great shading dependability amid preparing. Moreover, a great hygienic status, lack of bias in taste and smell, and lawful prerequisites should be taken into account [20].

Red Beet Powder, concentrates, and juices of red beet are the common applications for purposes of shading approved in North America and Europe and still speak to the sole origin of betalains utilized economically. Beginning in the mid-1970s, Von Elbe and collaborators reported the exploration actions of betalains that strengthened the applications of red beet. Normally, entire peeled beets are prepared; evacuation of the peels causes loss of more than 30% of color. This is critical in light of the fact that the best polyphenol oxidase action, which is pernicious to betaxanthins as well as betacyanins, is situated in the fruit covering. Unfavorable activities of enzymes can be made inactive via an instrument displayed by past blanching [21]. Basically the hydroxylating and oxidizing potentials of polyphenol oxidase require mono- or diphenolic cores once in a while present in both betacyanins and betaxanthins and strictly when β-glucosidase movement caused hydrolysis. Subsequently, for degradation of betalain via enzymatic activity, a coordinated activity of glucoside-separating chemicals such as peroxidases and polyphenol oxidases is essential. By and large, little beets are favored in light of the fact that they collect higher betalain fixations. Generally, the tissue comminution is performed by processing, trailed by fermentation of the subsequent pound via expansion of citrus extract until the pH is maintained at 4 [22]. Bringing down the pH blocks polyphenol oxidases potential while peroxidases might in any case be dynamic until the sifted juice is warmed at >75 °C. In this way, colors will be less seriously influenced. Nonetheless chamber filtration, ultrafiltration has been effectively connected. Later methodologies utilizing beat electric fields for extraction of color have not yet entered mechanical rehearsal. Other than mechanically arranged work concentrated on solidness approaches amid preparing, a broad-rearing project started in the 1980s brought about a shading yield change of 200% [23].

For nourishment of coloring purposes, high pigment-low solid beets were proposed. Prominently, nearly little studies are one to misusing elective sustenance origins. This is astonishing since the arrangements of red beet are suppressed with high nitrate. Then again, red beet is financially attractive in light of its yearly production of 50–70 tons for each hectare comprising of 40–200 mg betanin/100 g that is not obtained by any other crop of betalains crop [24] (Fig. 9.2).

In physical unselective extraction in view of oil or water completed by focus warming is permitted, taking after the late pattern for clean labelled sustenance. The obtained concentrates of fruit or vegetable might be connected is to bolster the specific food coloring attributes. Regular smell and tastes of the shading crop sources describe them. These hued concentrates are named as fixings, e.g., extricates of red beets [25]. Then again, if the fermentation of color concentrate is carried out with molds or yeasts to evacuate sugars for accomplishing a greater tinctorial quality after five–seven-fold fixation till 65°Bx is achieved, the subsequent item is viewed as a characteristic pigment. Same is the case of shading arrangements that have been denitrified beforehand. Amid aging, the betacyanins ended up being stable than the betaxanthins, which is thought to be because of their warm dependability as opposed to various propensities of colors toward microbial degradation. Other than these biological apparatuses, beet concentrates might likewise be filtered by section chromatographic procedures [26]. After evacuation of salts, phenolics, and sugars, the nature-determined coloring readiness will, nonetheless, require E number naming.

Amaranth For leaf and grain amaranth, an extraordinary potential has been estimated for ornamental as well as nutritious purposes. The color of leaves of amaranth is utilized as nourishment shading as a part of Northwestern Argentina and Bolivia for mixed drinks, in Ecuador for various types of sustenance, and in the Southwest USA and Mexico as a colorant for maize batter. It is approved for the characteristic color production in China. In 1996, reproduction and determination program originated in China in which 388 genotypes of 8 different genera and 37 species were concentrated on. As compared to wild species, higher color substances were observed in developed species, and aggregate betacyanin substance ranges from 46 to 199 mg/100 g fresh material of the plant [27]. Because of greater extent of acylated betacyanins, an agreeable stability was accomplished in splash dried arrangements. In view of the more extensive versatility of amaranth plants contrasted with red beets, the previous were suggested as normal sustenance colorants as novel sources. Then again, saponins adding up to 0.1% of dry matter and dopamine substance of around 6 mg/g new weight should be precisely viewed as before their utilization in nourishments [28].

Prickly plant Pear *Opuntia* sp. involving cactus pear fruit are the center of the generation of betalain-based shading arrangements to broaden the limited tone scope of red beet arrangements since 1998. Its chromatic properties are seemed to be equivalent to red beet. Distinctive colors ought to be achievable together with orange fruits. A procedure was developed for *O. ficus-indica* cv. "Gialla" to produce a yellow-orange cactus pear juice from it that was stretched out to shower coloring focuses and dried powders on a semi-mechanical level [29]. Lab-scale preparation of a red-purple cactus pear has been accounted for, whereas *O. stricta* was used to obtain a dark red pigment concentrate as a violent pigment, commercial anthocyanic extricates, and red beets. Though a 100 g natural product produces15–80 mg yield, high yields of 100 mg can be obtained via 100 g of new crossbreeds. Taking into account these facts, it can be said that cactus pears can serve as a source of yellow-orange, red, and purple pigmentation, and their utilization will have a splendid future [27].

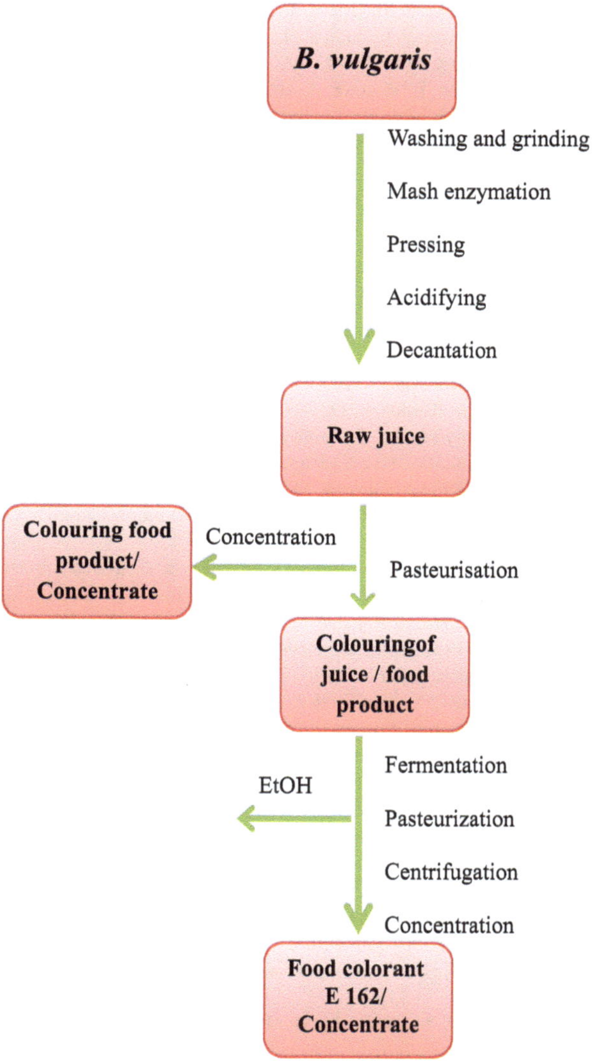

Fig. 9.2 Generation of food color and coloring foodstuff from *B. vulgaris* (red beet)

References

1. D'Archivio, M., Filesi, C., Varì, R., Scazzocchio, B., & Masella, R. (2010). Bioavailability of the polyphenols: Status and controversies. *International Journal of Molecular Sciences, 11*(4), 1321–1342.
2. Pavokovic, D., & Krsnik-Rasol, M. (2011). Complex biochemistry and biotechnological production of betalains. *Food Technology and Biotechnology, 49*(2), 145.
3. Clifford, T., Howatson, G., West, D. J., & Stevenson, E. J. (2015). The potential benefits of red beetroot supplementation in health and disease. *Nutrients, 7*(4), 2801–2822.

4. Livrea, M. A., & Tesoriere, L. (2006). Health benefits and bioactive components of the fruits from Opuntia ficus-indica [L.] Mill. *Journal of the Professional Association for Cactus Development, 8*(1), 73–90.
5. Esatbeyoglu, T., Wagner, A. E., Schini-Kerth, V. B., & Rimbach, G. (2015). Betanin—A food colorant with biological activity. *Molecular Nutrition & Food Research, 59*(1), 36–47.
6. Gandía-Herrero, F., Escribano, J., & García-Carmona, F. (2016). Biological activities of plant pigments betalains. *Critical Reviews in Food Science and Nutrition, 56*(6), 937–945.
7. Neelwarne, B. (Ed.). (2012). *Red beet biotechnology: Food and pharmaceutical applications*. New York: Springer Science & Business Media.
8. Pavokovic, D., & Krsnik-Rasol, M. (2011). Complex biochemistry and biotechnological production of betalains. *Food Technology and Biotechnology, 49*(2), 145.
9. Khan, M. I. (2016). Plant betalains: Safety, antioxidant activity, clinical efficacy, and bioavailability. *Comprehensive Reviews in Food Science and Food Safety, 15*(2), 316–330.
10. Azeredo, H. (2009). Betalains: Properties, sources, applications, and stability–a review. *International Journal of Food Science & Technology, 44*(12), 2365–2376.
11. Walkowiak-Tomczak, D. (2002). Microbiological denitrification of red beet juice. *European Food Research and Technology, 215*(5), 401–406.
12. Schliemann, W., Cai, Y., Degenkolb, T., Schmidt, J., & Corke, H. (2001). Betalains of Celosia argentea. *Phytochemistry, 58*(1), 159–165.
13. Khan, M. I., & Giridhar, P. (2015). Plant betalains: Chemistry and biochemistry. *Phytochemistry, 117*, 267–295.
14. Delgado-Vargas, F., Jiménez, A. R., & Paredes-López, O. (2000) Natural Pigments: Carotenoids, Anthocyanins, and Betalains—Characteristics, Biosynthesis, Processing, and Stability, *Critical Reviews in Food Science and Nutrition, 40*, 173–289
15. Pourrat, H., Lejeune, B., Regerat, F., & Pourrat, A. (1983). Purification of red beetroot dye by fermentation. *Biotechnology Letters, 5*(6), 381–384.
16. Vergeront, T. A., Von Elbe, J. H., & Amundson, C. H. (1980). Large-scale isolation of betalains by gel filtration. *Process Biochemistry, 15*(2), 15.
17. Brenner, D. M., Baltensperger, D. D., Kulakow, P. A., Lehmann, J. W., Myers, R. L., Slabbert, M. M., & Sleugh, B. B. (2010). Genetic resources and breeding of Amaranthus. *Plant Breeding Reviews, 19*, 227–285.
18. Teutonico, R. A., & Knorr, D. (1985). Amaranth: Composition, properties, and applications of a rediscovered food crop. *Food Technology (USA), 39*, 49.
19. Singh, G., Kawatra, A., & Sehgal, S. (2001). Nutritional composition of selected green leafy vegetables, herbs and carrots. *Plant Foods for Human Nutrition (Formerly Qualitas Plantarum), 56*(4), 359–364.
20. Socaciu, C. (Ed.). (2007). *Food colorants: Chemical and functional properties*. Boca Raton: CRC Press.
21. Cai, Y., Sun, M., & Corke, H. (1998). Colorant properties and stability of Amaranthus betacyanin pigments. *Journal of Agricultural and Food Chemistry, 46*(11), 4491–4495.
22. Cai, Y. Z., Sun, M., & Corke, H. (2005). Characterization and application of betalain pigments from plants of the Amaranthaceae. *Trends in Food Science & Technology, 16*(9), 370–376.
23. Cai, Y., Sun, M., & Corke, H. (2001). Identification and distribution of simple and acylated betacyanins in the Amaranthaceae. *Journal of Agricultural and Food Chemistry, 49*(4), 1971–1978.
24. Cai, Y., Sun, M., Wu, H., Huang, R., & Corke, H. (1998). Characterization and quantification of betacyanin pigments from diverse Amaranthus species. *Journal of Agricultural and Food Chemistry, 46*(6), 2063–2070.
25. Cai, Y. Z., & Corke, H. (2000). Production and properties of spray-dried Amaranthus Betacyanin pigments. *Journal of Food Science, 65*(7), 1248–1252.
26. Oleszek, W., Junkuszew, M., & Stochmal, A. (1999). Determination and toxicity of saponins from Amaranthus cruentus seeds. *Journal of Agricultural and Food Chemistry, 47*(9), 3685–3687.

27. Stintzing, F. C., Schieber, A., & Carle, R. (2003). Evaluation of colour properties and chemical quality parameters of cactus juices. *European Food Research and Technology, 216*(4), 303–311.
28. Stintzing, F. C., Schieber, A., & Carle, R. (1999). Amino acid composition and betaxanthin formation in fruits from Opuntia ficus-indica. *Planta Medica, 65*(7), 632–635.
29. Stintzing, F. C., Schieber, A., & Carle, R. (2001). Phytochemical and nutritional significance of cactus pear. *European Food Research and Technology, 212*(4), 396–407.
30. Pavokovic, D., Rusak, G., Besendorfer, V., & Krsnik-Rasol, M. (2009). Light-dependent betanin production by transformed cells of sugar beet. *Food Technology and Biotechnology, 47*, 153–158.
31. Stintzing, F.C., & Carle, R. 2008. Analysis of betalains. *In Food colorants: Chemical and functional properties*, ed. C. Socaciu, 507–520. Boca Raton: CRC Press.
32. Guevara, J.C., Suassuna, p. & Felker, P. (2009). Opuntia forage production systems: status and prospects for rangeland application. *Rangeland Ecology and Management, 62*, 428–434.
33. Bellec, F. L., Vaillant, F., & Imbert, E. (2006). Pitahaya (Hylocereus spp.): a new fruit crop, a market with a future. *Fruits, 61*, 237–250.
34. Vaillant, F., Pérez, A., Dávila, I., Dornier, M., & Reynes, M. (2005). Colorant and antioxidant properties of red-purple pitahaya *(Hylocereus sp.). Fruits, 60*, 1–10.
35. Frank, T., Stintzing, F. C., Carle, R., Bitsch, I., Quaas, D., Gabriele, S., Bitsch, R., & Netzel, M. (2005). Urinary pharmacokinetics of betalains following consumption of red beet juice in healthy humans. *Pharmacological Research, 52*, 290–297

Chapter 10
Processing of Betalains

10.1 Introduction

The existing lifestyle derived the human needs to the enhanced consumption of processed food. The concept of food preservation is very traditional. Since ancient times, drying and preservation by salting are the most popular methods which prevent food from biodegradation caused by bacteria, thus resulting in increased shelf life. The food processing generally extends the storage period of food by applying various preservation techniques. This avoids the biochemical as well as microbiological changes in food. In contemporary practice, processing of food enhances the nutritional value and variety of food incorporating organoleptic effect and nutritional quality. Food obtained from plants is generally processed through thermal methods to increase shelf life, flavor and enhances the scrumptiousness. The overall purpose is to make the food safe and available around the year for human consumption.

Betalains undergo several changes during its processing. Increasing the yield of betalain by processing is required as a prior condition. Extracted betalains are sensitive to pH, moisture, light revelation, oxidation, metal ions, temperatures, and enzymatic actions. However, this sensitivity retards food coloring property of betalains. On the other hand, processing alters the betalain contents, food color, and its antioxidant power. Processing of food includes dehydration, canning, fermentation, freezing, freeze-drying, pickling, and pasteurization. The two methods employed for betalain storage are thermal processing and nonthermal processing.

10.2 Thermal Processing

Thermal processing includes microwaving, boiling, and roasting that decreases the stability of betalains. Some disadvantages are also associated with thermal processing, i.e., betalain gets degraded under high temperature. By boiling and roasting, the

E. Akbar Hussain et al., *Betalains: Biomolecular Aspects*,
https://doi.org/10.1007/978-3-319-95624-4_10

yields of betacyanins and betaxanthins are affected [1]. Betanin undergoes degradation by hydrolytic cleavage to produce betalamic acid and cyclo-dopa 5-*O*-glucoside as biosynthetic precursors. It is also regenerated up to some extent by recondensation of the hydrolyzed products through which its color is recovered after keeping the heated extracts in cold atmosphere. Thermal exposure causes isomerization and decarboxylation of betanin to produce its C-15-stereoisomer isobetanin and descarboxybetanin, respectively [2].

Microwave treatment Betaxanthin contents tend to increase by high treatment conditions under microwave irradiations like 900 watts for 30 s and 1800 watts for 30 s. Betaxanthin contents increase slightly by 7% during 900 W and by 19% during 1800 W. By increasing temperature, betacyanin contents also increase. Microwave heating affects the natural colors of fruit purees. In order to avoid betalain degradation caused by enzymatic conditions, enzyme is inactivated by treatment with low heat of 70 °C for 2 min. Stability of betacyanins is improved by acidification. Acidification also avoids oxidation caused by enzyme polyphenol oxidases. Red beets contain endogenous enzymes like β-glucosidases, polypheoloxidases, and peroxidases. These enzymes cause color loss and degradation of betalains if not inactivated properly.

Boiling and Roasting Betacyanin contents decrease by changing the time in boiling treatment; by increasing the time from 60, 120, and 180 s, the betacyanin contents are decreased by 6%, 22%, and 51%, respectively, and betaxanthin contents are decreased by 18%, 23%, and 33%, respectively. By treating with 110, 115, and 125 °C for 30 min, betacyanin of beets tend to decrease by 24%, 62%, and 81%, and its betaxanthins contents are decreased by 13%, 60%, and 73%, respectively. These results reveal the processing temperatures effect on betacyanins and betaxanthins. Betaxanthin contents tend to increase 7% after roasting for 5 min.

10.3 Nonthermal Processing

Vacuum Treatment Nonthermal treatment such as vacuuming prevents the degradation of betalains. Vacuum treatment involves available oxygen removal. It also results decrease in pigment degradation under low oxygen levels than under atmospheric oxygen. Degraded pigments are recovered partially under low oxygen levels. Betalains are oxidation-sensitive, influencing their color stability. Moreover, temperature is one of the most important factors that affect the stability of betalains.

In vacuum-dried red beet tissue extracts, betacyanin contents are increased from 7 to 8.5 mg/L, and betaxanthin contents are increased from 5.2 to 5.8 mg/L as compared to the samples that were treated by non-vacuum processing. In controlled setup, betacyanin contents are increased by 20%, and betaxanthin contents are increased by 12%.

10.4 Effects of Processing on Antioxidant Activity

Processing increases the antioxidant power of betalain. This increase is due to degradation of vitamin contents, as betalains have higher antioxidant activity than that of ascorbic acid. High antioxidant activity of treated samples revealed that it depends not only on the presence of betalains but also on other polyphenols which in turn also increased during the processing treatments, except for treatments with 450 W for 20 s, 900 W for 10 s, and 1800 W for 10 s and also during roasting for 1 min. It was found that in complete vacuum treatment, the antioxidant activity was increased up to 15%.

Microwave treatments increase antioxidant activity up to twofolds. Boiling and roasting increased antioxidant activity up to threefolds. After processing, antioxidant activity of red beet is increased due to minor initial treatments at high temperatures. It is related to their various vitamin contents and phenolic compounds that operate in interactive manner and process sensitively. Cooking enhances antioxidant activity dramatically in tropical green leafy vegetables [3].

Processing problems like degradation of betalains and recovery of pigments are economically essential in food items and must be solved in order to obtain betalain-based dyes. Commercial betalain effectiveness is dependent on a continuous availability of highly pigmented sources and the use of cold and modified storage atmospheres except for processing, efficient enzymatic control, handling practices, extraction procedures, purification, concentration, and finishing operations like freeze, spray, and vacuum drying [4].

References

1. Ravichandran, K., Thaw Saw, M., Adel, A. A., Mohdaly, M. M., Anja, K., Heidi, R., et al. (2013). Impact of processing of red beet on betalain content and antioxidant activity. *Food Research International, 50*, 670–675.
2. Florian, C. S., & Reinhold, C. (2007). Betalains–emerging prospects for food scientists. *Trends in Food Science & Technology, 18*(10), 514–525.
3. Adefegha, S. A., & Oboh, G. (2009). Cooking enhances the antioxidant properties of some tropical green leafy vegetables. *African Journal of Biotechnology, 10*(4), 632–639.
4. Delgado-Vargas, F., Jimenez, A. R., & Paredes-Lopez, O. (2000). Natural pigments: Carotenoids, anthocyanins, and betalains — Characteristics, biosynthesis, processing, and stability. *Critical Reviews in Food Science and Nutrition, 40*(3), 173–289.

MIX
Papier aus verantwortungsvollen Quellen
Paper from responsible sources
FSC® C105338

If you have any concerns about our products,
you can contact us on
ProductSafety@springernature.com

In case Publisher is established outside the EU,
the EU authorized representative is:
Springer Nature Customer Service Center GmbH
Europaplatz 3, 69115 Heidelberg, Germany

Printed by Libri Plureos GmbH
in Hamburg, Germany